L: 50 Objects, Stories & Discoveries from The Linnean Society of London

Second Edition

With a foreword by Sir David Attenborough OM

The LINNEAN SOCIETY of London

First published by The Linnean Society of London in 2020.

Second edition published in 2025.

ISBN 978-0-9935510-2-4

British Library Cataloguing-in-Publication data:
A catalogue record for this book is available from the British Library.

Photography and design: Leonie Berwick
Additional photography: Andrea Deneau

Printed and bound in the UK by Bishops Printers

Printed on Silk (FSC) Mixed Credit stock using vegetable-based inks with ISO14001 accreditation.

The Linnean Society of London
Burlington House, Piccadilly
London W1J 0BF UK
www.linnean.org

UK registered charity no. 220509

L: 50 Objects, Stories & Discoveries from The Linnean Society of London

Compiled & edited by Leonie Berwick & Isabelle Charmantier

Contributors

George Beccaloni (GWB)
Will Beharrell (WB)
Glenn Benson (GB)
Leonie Berwick (LB)
Ranit Bhuyan (RB)
Tim R. Birkhead (TRB)
Rich Boden (RBo)
Gunnar Broberg (GBr)
Isabelle Charmantier (IC)
Elaine Charwat (EC)
James T. Costa (JTC)
Anne Courtney (AC)
Oliver Crimmen (OC)
John David (JD)
Gina Douglas (GD)
John Edmondson (JE)
Brian J. Ford (BJF)
Roger Gaskell (RG)
Taylor Harwood (TH)
Paul Henderson (PH)
Stephanie Holt (SH)
Elena Isayev (EI)
Mark James (MJ)
Charlie Jarvis (CJ)
Tom Kennett (TK)
Julie Kim (JK)
David Lowther (DL)
James Maclaine (JM)
Liz McGow (LM)
Vida Milovanovic (VM)
Staffan Müller-Wille (SMW)
Henry Noltie (HN)
Eva Nyström (EN)
Lynn Parker (LP)
Tim Robinson (TR)
Mark A. Spencer (MAS)
Mark Watson (MW)
Kristen Wellborn (KW)
Rebecca J. Wilson (RJW)

Contents

Foreword

You might walk past the Linnean Society on London's Piccadilly and be unaware of all the astonishing treasures that lie within.

From the world's first cultured spherical pearls to the very microscope on which the concept of the cell nucleus was established—the Society has been a home to natural science, history and art for centuries.

Founded in 1788 upon the purchase of the collections of the Swedish naturalist Carl Linnaeus (or Carl von Linné, 1707–1778), the Linnean Society has been caretaker of important biological specimens, rare books, manuscripts, artworks and artefacts from the 18th century to the present.

As a meeting place for anyone interested in the natural world, the Society has witnessed some extraordinary events, not least the first reading of Charles Darwin and Alfred Russel Wallace's theories of evolution by natural selection in 1858, which profoundly changed the way in which we see our world. And by November 1904, the Society was one of the earliest to elect women as members, with numerous trailblazers publishing papers and attending meetings, like renowned plant morphologist and anatomist Agnes Arber. Many of their portraits, alongside images of nature from the collections, adorn the walls of the Society's historic home.

Like their 18th-century counterparts, the study and stewardship of—as well as passion for—the natural world is what drives the Society's global membership today. All of this is what makes the Linnean Society so special.

This book uncovers just a fraction of the Society's many treasures and I hope it will inspire you to join us in cultivating a world where nature is understood, valued and protected. At a time when the natural world is under threat, connection through collections, art and accessible research is vital in inspiring its future protection.

Sir David Attenborough OM, Vice Patron of the Linnean Society

Introduction

Isabelle Charmantier, Head of Collections, The Linnean Society of London

'L'

It is probable that the Swedish naturalist Carl Linnaeus (1707-1778—Top Left), famed for his classification of the natural world, never thought that his legacy would live on through his surname, appearing after the plethora of binomial names he had coined, mostly through the use of his initial 'L.'. Devised in the 1750s, Linnaeus' binomial, or 'two-name' nomenclature moved away from long, cumbersome descriptions in Latin and simplified them to binomials: 'genus' and 'species'. So, a sunflower is *Helianthus annuus* L., an oak is *Quercus robur* L., and we are *Homo sapiens* (Linnaeus, 1753 and 1758). This made it much easier for naturalists to communicate and know they were referring to the same plant or animal. It was nothing short of revolutionary, and is still in use today.

The botanist James Edward Smith (1759-1828—Bottom Left), who in 1784 bought the Linnaean Collections on the advice of Sir Joseph Banks (1743-1820), founded the Linnean Society in 1788 to 'place the authority of this collection, as far as possible, out of the reach of accident, [...and] to extend any information to be derived from it, not only to his own countrymen, but to his fellow-labourers in every quarter of the globe'. He spent the rest of his career translating some of Linnaeus' works, putting the biological collections in order, and opening them up to the scientific community. Papers on new species of plants and animals from around the world, new theories, and new discoveries were read and discussed at meetings attended by Fellows of the Linnean Society.

The legacy of this history is found in the rich collections of the Society (Opposite). At the heart of these are the unique Linnaean collections: Linnaeus' books, manuscripts, letters and biological specimens. As wonderful as they are, the Linnaean Collections represent just a small portion of the Society's holdings. Over 230 years of meetings, discussions, donated specimens, manuscripts, illustrations and artefacts have resulted in materials that are often just as astonishing and priceless as the Linnaean collections.

The Society's Collections team has long given tours, during which Fellows and members of the public are shown the Linnaean collections and the artwork dotted around the building. But many more treasures remain locked away behind the shelves of our library and archives. As these 'Treasures Tours' have grown in popularity over the years, with the number of visitors doubling in half a decade, it is time that the Linnean Society had its own 'Book of Treasures'.

If 'L' stands for Linnaeus, it also represents the Roman numeral 50. This book aims to showcase 50 of the Linnean Society's well-known and rarer treasures.

We called upon the expertise of our Fellows and curators to help research and write many of the sections that follow. Along the way, we have discovered more about the items we thought we knew so well, and unearthed some hidden

gems: a well-thumbed entrance token to the Amsterdam botanical garden (p. 16), the first image of a Red Panda to arrive on European shores (p. 70), and an enigmatic piece of wrapping from a British Museum mummy (p. 116) stand side-by-side with Linnaeus' iconic Lapland journey diary (p. 22) and Robert Brown's microscope (p. 72).

The items on the following pages are just a small taste of what the Society holds. We hope this book will entice our Fellows and researchers to further explore our irreplaceable and remarkable collections. And, by opening up our collections in *L: 50 Objects, Stories and Discoveries from The Linnean Society of London*, we invite you to discover these incredible treasures for yourselves.

This second edition has benefitted from new research on several of the items, as well as new acquisitions, which have only enhanced our understanding of our holdings. It is wonderful to open up our collections in this way, and we hope this new edition demonstrates how there is always something new to learn about the Linnean Society and its rich collections.

Early Modern

As Europe emerged from the Middle Ages, new worlds, knowledge and species were discovered and described through investigation, debate and the revolutionary medium of print. The Linnean Society's collections contain a number of fascinating items from this period.

De Urinis.

Q

Uoniam medicus est
artifex sensitiuus. ⁊ ꝑ
signa ī egritudinum
causas deueniat. opꝰ
est vt maxime aduer
tat in medicando ad
ꝑncipalia signoꝝ ge
nera. que q̃tuor a me-
dicis appellari consueuerunt. vt colligit̄
in summa Galieni in tegni. Quoꝝ p̄mū
sumitur a tumoribus. Scdm̄ a doloribus
Tercium ab opationibus. Quartū a sup
fluitatibus. Vez quia signa sumpta a su
perfluitatibus obijciunt̄ magis sensibus
nostris. magisq̄ sensitiua dicunt̄. Et cum
inter om̄ia vrina obtineat principatū de-
creui tractare de vrinis. ex dictis autorū
medicum: totis viribus congregare. Nō
vt aliquid noui me factuꝝ existimem: Sz
vt amorem dn̄oꝝ scolariū. quoꝝ filius sū
geram voluntati. Indigēt em̄ subtili cog-
nitione circa vrinarum species. Cum ma
xime vt referunt decipiuntur in ꝑtibꝰ coꝝ
Tria enim cum melius potero adducam
Primo regulas quasdam. quas hab; me
dicus tanq̄ pro suppositione. anteq̄ ali
quod iudicium faciat de vrinis. Secūdo
regl'as sumptas a coloribus. Tertio regu
las sumptas a contentis. Quantum er
go ad primum dimisso quid nominis vri
ne quod tangitur ab Egidio carmine pri
mo. descendam ad principale. Primo no
tandum quod ab vrina varia sumunt̄ sig
na. Primo a colore. Secūdo a substantia
Tercio a contentis. que varias dispositō
nes habent iudicare: Color enim arguit
super complexionem. sc; calidi ⁊ frigidi:
cum causetur a qualitatibus actiuis. scili
cet a caliditate ⁊ frigiditate Substantia
vero significat supra humores cum cause
tur a qualitatibus passiuis scilicet humi
ditate ⁊ siccitate. lic; per accidens possit
color causari a passiuis. vt ex permixtōne
rei humide. ⁊ sanguinis cū vrina. Contē
tum autem super dispositionem mēbroꝝ.
Ex quo sequitur immediate solutō ad du

bitationem c
arguit nisi su
rum alias int
patet ex dicti
cognoscere m
caliditatis. I
tice. Secund
tur ab Auice
sumptis ab v
⁊ alijs dispos
inferius. Qu
regule. Pr
na secunda p
quam medic
dicium facia
eam in vase
habente form
vrinale. vtme
torum. Tale
siue simili loc
tur ab Ysaac
um. Ignari n
sse iudicare d
⁊ ī alijs vasi
cognoscere s
uulsiones yp
litatez eiusd
cōpleram se
r̄gula sumpt
Et ab Auice
stans in vase
aere radioso.
sed in loco a
batur primo
vas impedi
sint causa dis
uientium po
ris habere iu
lis in fundo e
penetrantib
locus conten
do sunt nim
nimis terrest
media. aliqu
fra patebit d
regule pater.

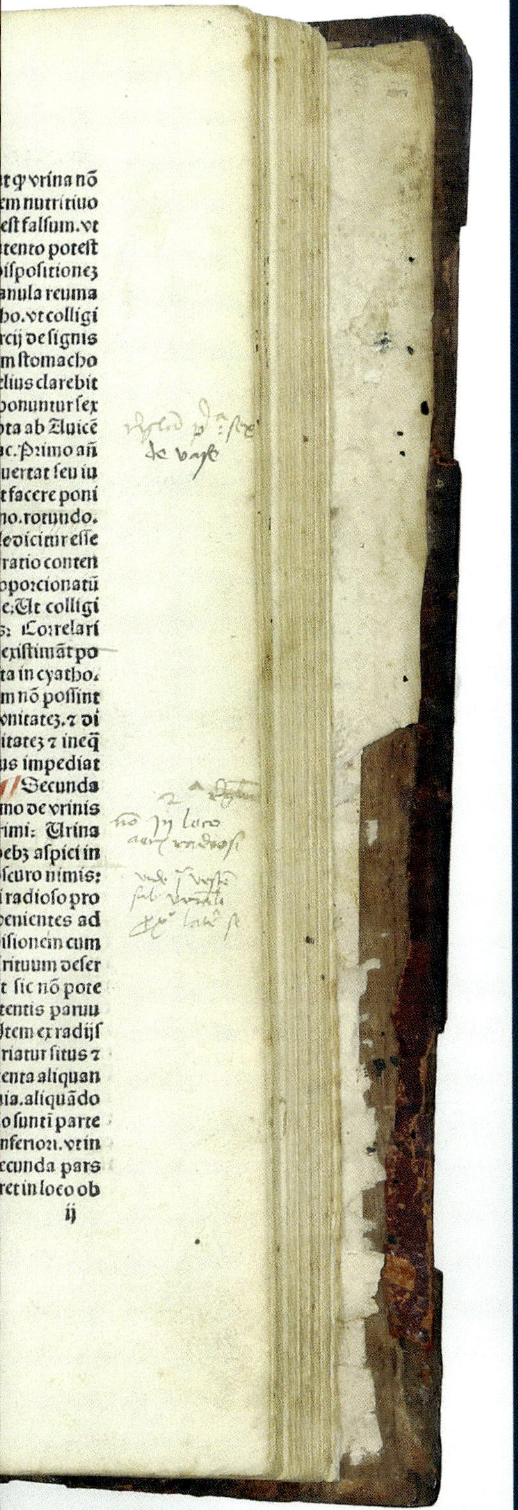

Ortus Sanitatis or *The Garden of Health*

1491

First published in Germany in 1485, the fascinating *Ortus Sanitatis* (or *The Garden of Health*) is a window into the late Middle Ages, when the natural world was thought to have been created by God for man's use. The book describes plants, animals and minerals that have medical properties, and their modes of preparation. Mythical creatures are also included, the pages filled with dragons, mermaids and other monsters. The last section, on the analysis of urine (the starting point for many medical diagnoses in the Medieval period), is illustrated by a woodcut showing medical practitioners in a shop examining phials of urine. Two children seem to be fighting in the foreground, perhaps afflicted by the choleric disposition that the physician is trying to diagnose. The charm of this wonderful book rests in its woodcut illustrations. Many of the plants, while stylised, are easily recognisable, and the animal and mineral woodcuts offer images of everyday life.

The Linnean Society actually holds four editions of *Ortus Sanitatis*, and the title pages indicate the successive readers to whom these books belonged; unknown, faded names that hint at the books' long histories, before coming into the hands of the Society.

Two of the four copies—the second edition (1491) and the fourth edition (1499)—belonged to Carl Linnaeus. The 1491 edition is particularly impressive, being one of the bulkiest books in Linnaeus' personal library, with numerous annotations throughout and some hand-coloured illustrations. The two remaining editions are from 1497 (the first to figure a woodcut of the human skeleton) and 1511 (the first Italian edition); both were bought by the Society during World War II, on 22 September 1942. **IC**

A Botanical Token of Esteem

1684

The Society holds within its eclectic collections a small, well-used brass token, dated 1684, which was appropriately donated to the Society in 1895 by physician and Fellow Dr John Braxton Hicks (1823–1897). It allowed the bearer entry into the Hortus Medicus (the botanic garden of the University of Amsterdam). The botanical garden was established in the 1630s, and by 1661 it had over 1,300 plant species listed in its catalogue. In 1665 the city of Amsterdam went through a major phase of expansion and the gardens needed a new home. The collection of medicinal plants was temporarily housed elsewhere before the Hortus Medicus (renamed Hortus Botanicus) was finally relocated to its current site in central Amsterdam in 1684.

Access to the new garden was tightly controlled, and entrance fees were charged. Doctors and apothecaries wishing to study the medicinal plants had to pay for official admittance to the gardens, which would also allow attendance of the professor of botany's weekly lectures. Brass tokens like the one shown here were issued to confirm that payment had been made.

The obverse of the token shows a vase of healthy flowers (the central bloom being a tulip), with a blank space below it on which to inscribe the initials of the token's owner. The reverse could not be more different, with an ominous image of a skeleton with a scythe, the figure resting its bony hand on an hourglass, alluding to the passing of time and human mortality. In this context however, the hourglass is also a visual metaphor that life can be extended through the actions of a well-informed physician; hourglasses can of course be reversed. While our ultimate fate is illustrated by the tomb, the long grass and tulips that surround it serve as a reminder that life goes on. **GB**

18th Century

The Linnean Society's 18th-century collections revolve around Carl Linnaeus, the naturalist whose life (1707–1778) spans nearly the whole of the Enlightenment. At the end of a century of insatiable curiosity, the foundation of the Society in 1788 allowed Fellows to use the Linnaean collections to deepen their understanding of the natural world.

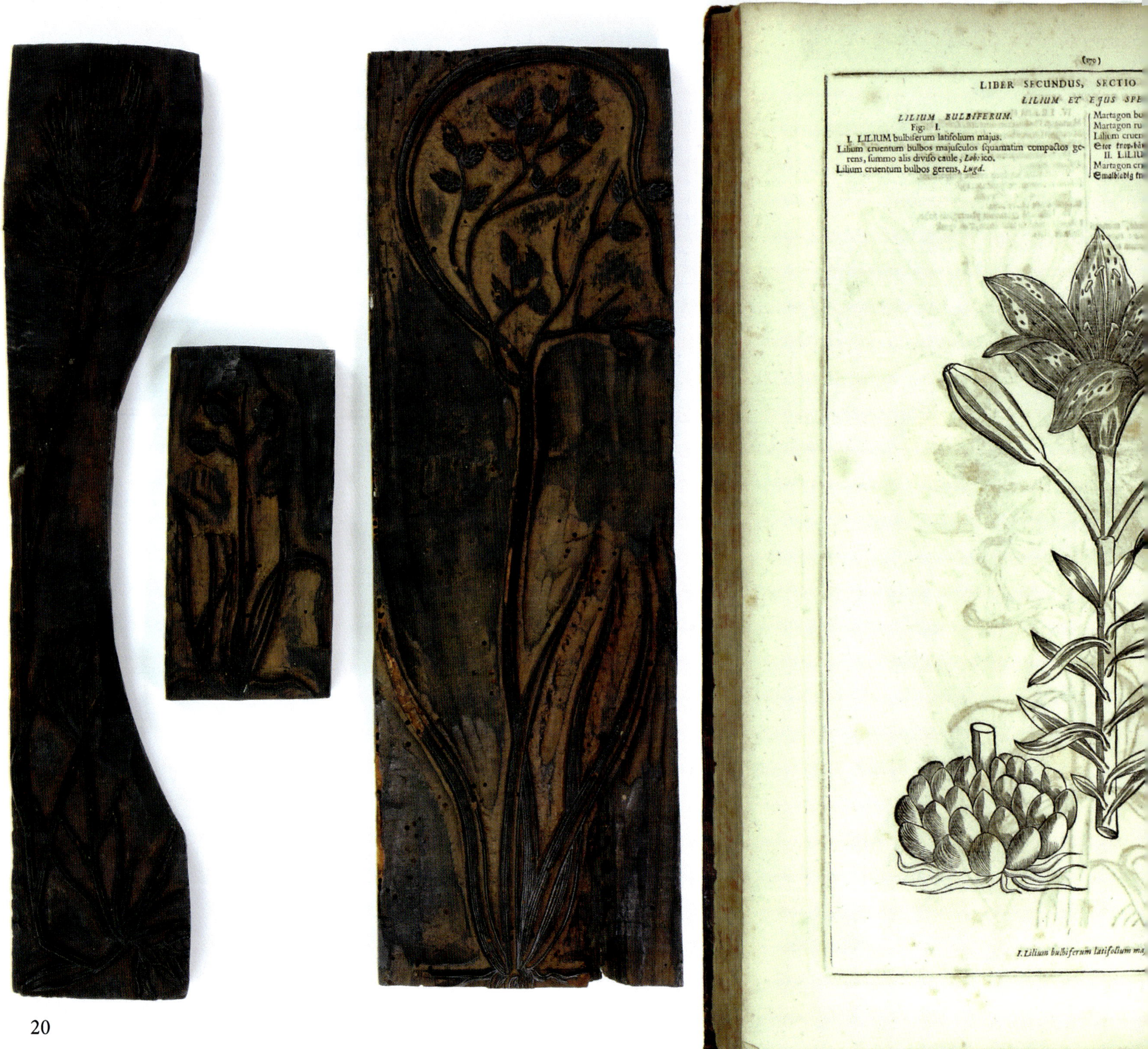
(170)
LIBER SECUNDUS, SECTIO
LILIUM BULBIFERUM.
Fig. I.
I. LILIUM bulbiferum latifolium majus.
Lilium cruentum bulbos majusculos squamatim compactos ge-rens, summo alis diviso caule, *Lobelio.*
Lilium cruentum bulbos gerens, *Lugd.*
I. Lilium bulbiferum latifolium ma,

Olof Rudbeck's Woodblocks

1702

Only two copies of the first volume of Olof Rudbeck's (1630–1702) *Campi Elysii* survived the great fire in Uppsala in May 1702; both are in Sweden. Remarkably, the Society has 25 of the woodblocks from which the illustrations were printed—rare survivors. The founder of the first botanical garden in Sweden in 1655, Rudbeck was appointed Professor of Medicine at the University in 1660: his anatomy theatre in the University, built in 1663, can still be visited.

Rudbeck worked on the *Campi Elysii* for most of his life, intending to illustrate all known plants full size in 10 volumes with 10,000 to 11,000 illustrations. Most of the 3,200 woodblocks already cut and copies of the first two volumes—as well as his manuscripts—were lost in the fire. Heartbroken, Rudbeck died a few months later. Olof Rudbeck the younger (1660–1740) succeeded his father as Professor of Medicine and after his death, his pupil Carl Linnaeus acquired the remaining blocks, which then passed to the Linnean Society. Still of taxonomic value nearly 100 years later, James Edward Smith used the original blocks to print *Reliquiae Rudbeckianae* (1798), where he included Linnaean nomenclature.

The Linnaean collections include one of a handful of surviving copies of he second volume of *Campi Elysii* (1701). Illustrating tulips and lilies, it shows some of the largest and most flamboyant plant pictures ever published in woodcut. The surviving woodblocks used to print the first volume (1702) illustrate graceful grasses and rushes, the smaller ones cut on boxwood, the larger ones on pearwood. In the preparation of the drawings and woodblocks, Rudbeck made use of the skills of his students and his children, including his daughters Johanna Christina and Wendela whose initials appear on some of the woodcuts. **RG & GBr**

Carl Linnaeus' Lapland Journal

1732

In the summer of 1732, Swedish naturalist Carl Linnaeus journeyed through Sápmi, a region in Northern Scandinavia known to him (and most English speakers today) as Lapland. During his trip, he composed a travel diary that celebrated the healthy, nomadic life of the Sámi people, but also contained much information that was supposed to further the colonisation of the region. Linnaeus has since been hailed as a pioneer of modern ethnography and ecology, with a keen eye for interactions between people, places and organisms.

Linnaeus' account references many ancient myths and legends—furies and chained maidens are woven into a narrative that makes it difficult at times to discern the fictional from the factual. But Linnaeus also relied on local knowledge, provided by people who helped him find his way: guides and servants, settlers and traders, midwives and wise women, priests and physicians, hunters and reindeer herders. While Linnaeus did contribute to an image of Lapland as a desolate region, waiting to be exploited for its natural resources, his journal also documents how locals knew precisely how to make a living there.

The Linnean Society holds the original manuscript of Linnaeus' 'Laplandic Journey' (*Iter Lapponicum*). Although he skillfully used his Lapland experience to promote himself to international fame, he never published the diary in its entirety. It first appeared in an English translation in 1811, commissioned by the Society's founder, James Edward Smith. Scores of scholars, scientists, writers and artists have since been re-visiting this manuscript. Linnaeus' writings on Lapland fuelled literary and artistic imaginations, and the diary continues to captivate and thrill readers to this day. **SMW & EI**

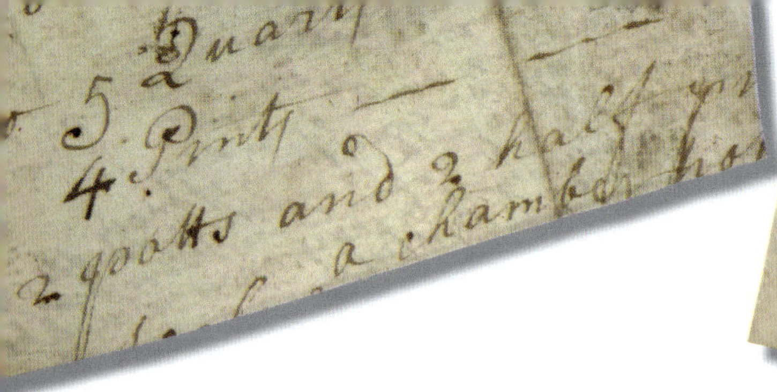

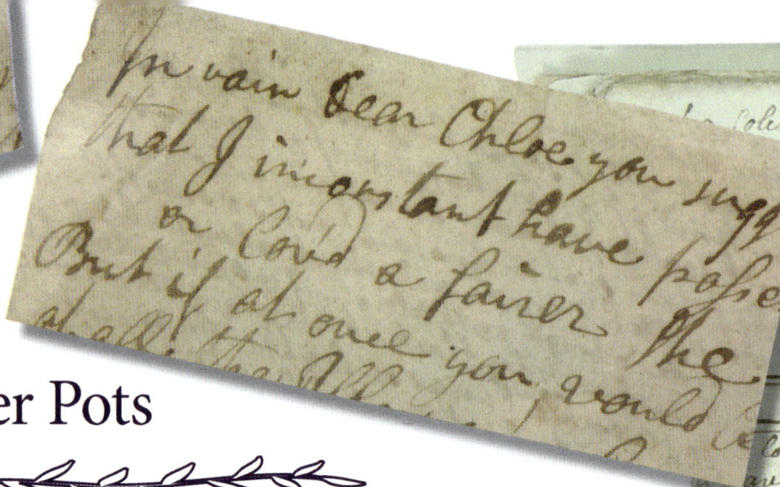

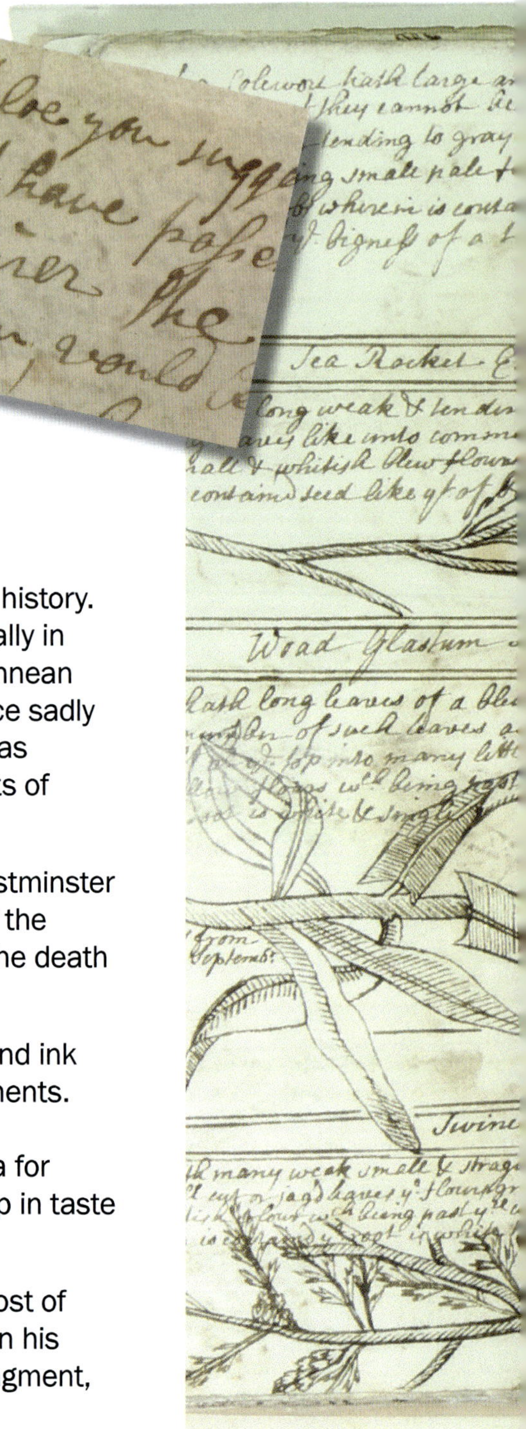

Plants, Poems & Chamber Pots

1733

One manuscript in the Society's archive collection has a fascinating yet turbulent history. The botanical notebook of the 18th-century physician Frank Nicholls was originally in the possession of Francis Druce (1873–1941), the Honorary Treasurer of the Linnean Society. Kept in his London flat, it was rescued after a bomb struck during the Blitz. Druce sadly lost his life, and the manuscript, which was salvaged from the remains of the building, was badly damaged by fire and water, and impossible to open. Fortunately, through the efforts of Conservator Janet Ashdown, it was lovingly restored in 2000.

Frank Nicholls (1699–1778) was born of Cornish parentage in London and attended Westminster School before studying medicine at Oxford. A Fellow of the Royal Society and member of the Royal College of Physicians, he was appointed physician to George II in 1753, following the death of Sir Hans Sloane.

The manuscript itself is a botanical notebook in two volumes, containing beautiful pen and ink drawings of British plants alongside detailed descriptions, with some further loose fragments. The descriptions can sometimes be wonderfully detailed, for example, mentioning the appearance, feel and locations of the plant, and Nicholls even sampled some of the flora for further research. The seeds of the 'Treacle Mustard Penny Cress', for example, are 'sharp in taste burning the tongue as doth mustard having the taste or savour of garlick behind'.

Even the reverse of the loose fragments are fascinating, as they have been cut from a host of interesting documents, including accounts (ordering a chamber pot!), letters to Nicholls in his capacity as a physician, and even part of a love poem addressed to 'dear Chloe'. One fragment, dated 1733, provides an approximate date for the notebook, which is undated. **LM**

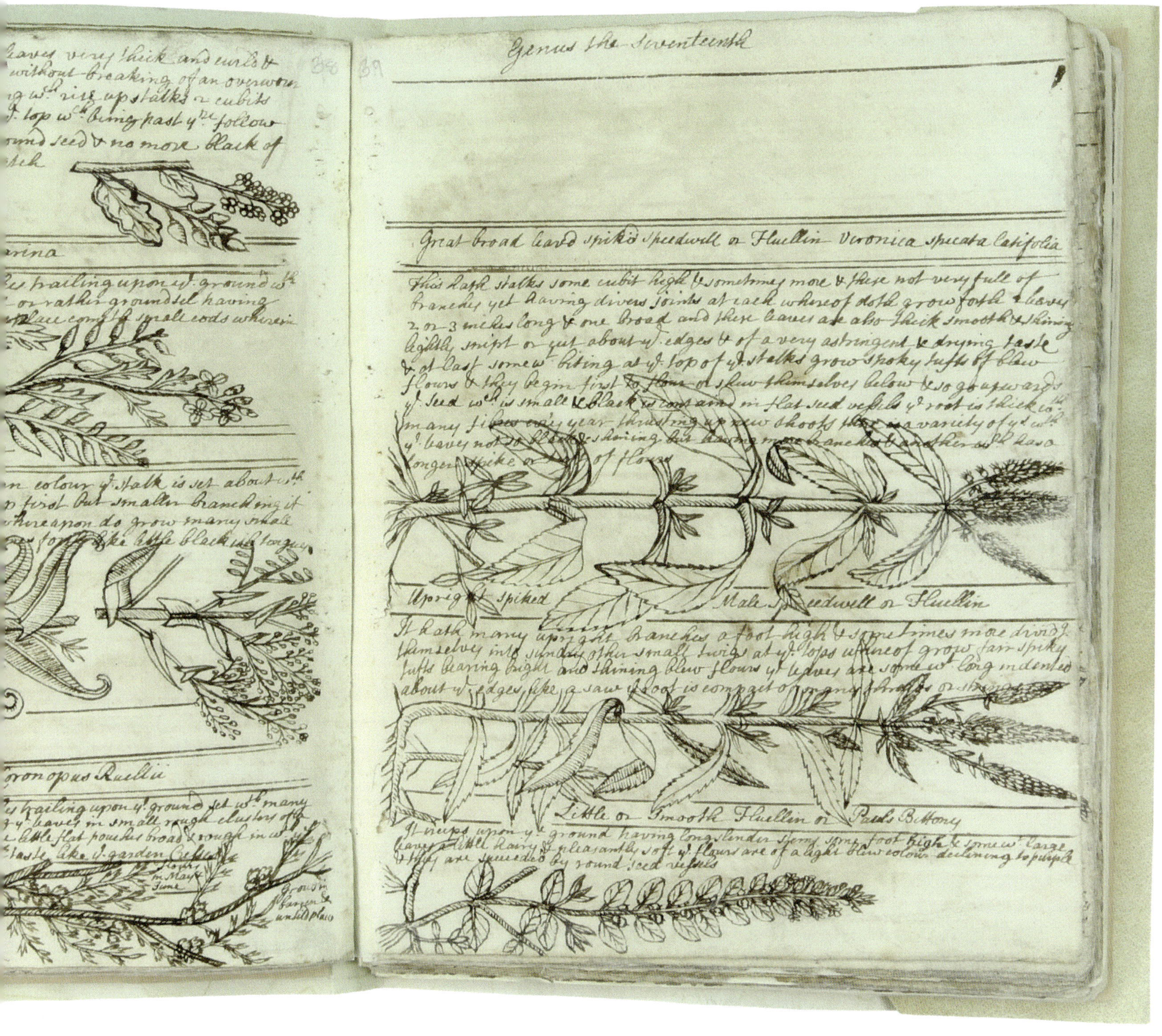

[left page, partial]
leaves very thick and curled & without breaking of an overworn [illegible] wch rise up stalks 2 cubits ye top wch being past ye[illegible] follow round seed & no more black of [illegible]tch

[illegible]rina

[illegible] trailing upon ye ground wth [illegible] or rather groundsel having [illegible] place come th small cods wherein

[illegible]n colour ye stalk is set about wth [illegible] first but smaller branching it [illegible] whereupon do grow many small [illegible] like little black [illegible] tongues

[illegible]oronopus Ruellii

[illegible] trailing upon ye ground set wth many [illegible] ye leaves in small rough clusters of [illegible] little flat pouches broad & rough in [illegible] taste like ye garden Cresses

Flours in May & June

Grows in barren & untild places

[right page]
Genus the Seventeenth

Great broad leav'd spik'd Speedwell or Fluellin Veronica spicata latifolia

This hath stalks some cubit high & sometimes more & these not very full of branches yet having divers joints at each whereof doth grow forth 2 leaves 2 or 3 inches long & one broad and these leaves are also thick smooth & shining lightly snipt or cut about ye edges & of a very astringent & drying taste & at last somewt biting at ye top of ye stalks grow spiky tufts of blew flours & they begin first to flow or shew themselves below & so go upwards ye seed wch is small & black is contained in flat seed vessels ye root is thick with many fibres every year thrusting up new shoots there is a variety of ye wch ye leaves not so [illegible] shining but having more branches & another wch has a longer spike or [illegible] of flours

Upright spiked Male Speedwell or Fluellin

It hath many upright branches a foot high & sometimes more dividing themselves into sundry other small twigs at ye tops whereof grow fair spiky tufts bearing bright and shining blew flours ye leaves are somewt long indented about ye edges like a saw ye root is compact of many threads or strings

Little or Smooth Fluellin or Pauls Bettony

It creeps upon ye ground having long slender stems some foot high & somewt large leaves a little hairy & pleasantly soft ye flours are of a light blew colour declining to purple & they are succeeded by round seed vessels

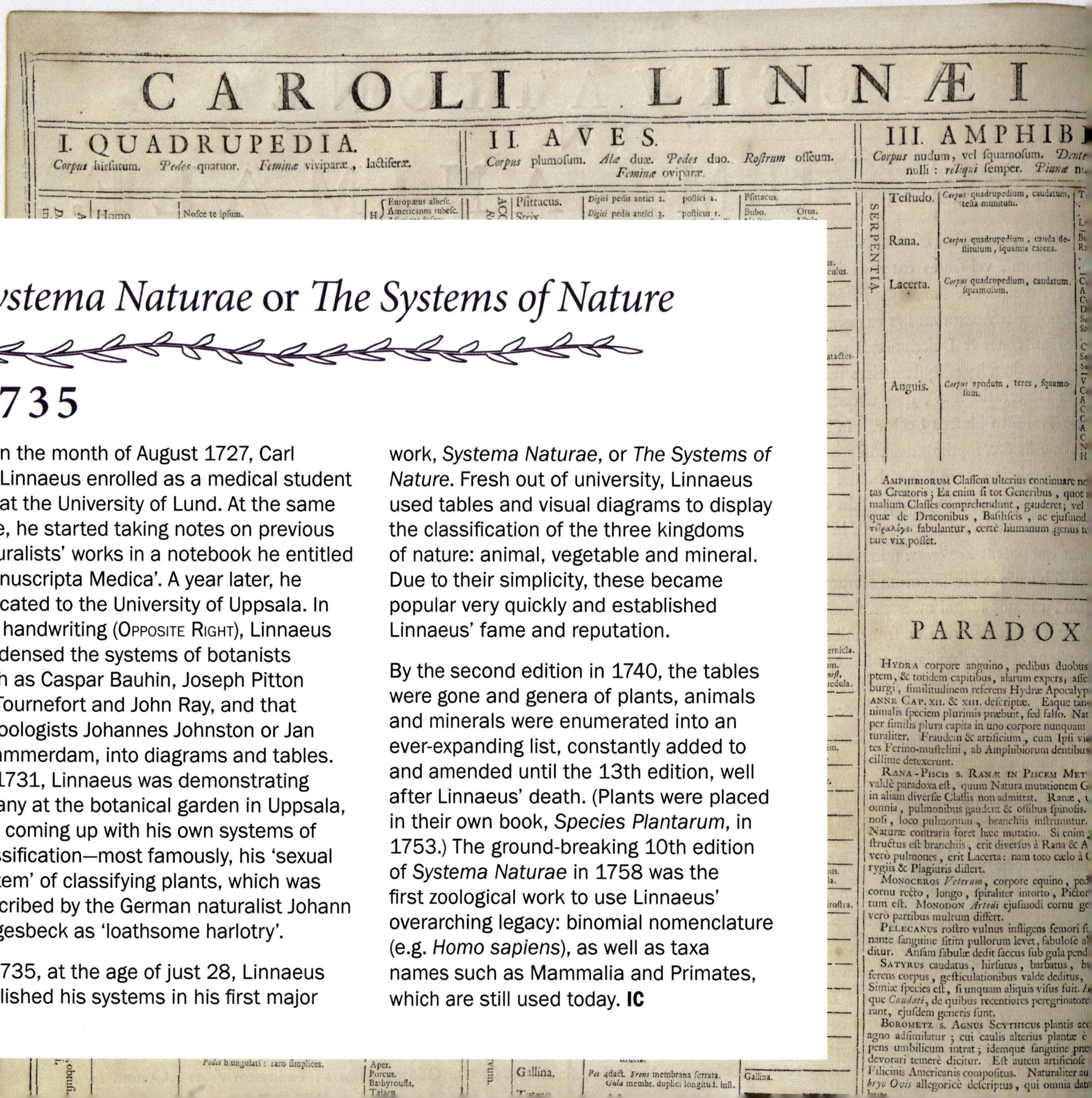

Systema Naturae or *The Systems of Nature*

1735

In the month of August 1727, Carl Linnaeus enrolled as a medical student at the University of Lund. At the same time, he started taking notes on previous naturalists' works in a notebook he entitled 'Manuscripta Medica'. A year later, he relocated to the University of Uppsala. In tiny handwriting (Opposite Right), Linnaeus condensed the systems of botanists such as Caspar Bauhin, Joseph Pitton de Tournefort and John Ray, and that of zoologists Johannes Johnston or Jan Swammerdam, into diagrams and tables. By 1731, Linnaeus was demonstrating botany at the botanical garden in Uppsala, and coming up with his own systems of classification—most famously, his 'sexual system' of classifying plants, which was described by the German naturalist Johann Siegesbeck as 'loathsome harlotry'.

In 1735, at the age of just 28, Linnaeus published his systems in his first major work, *Systema Naturae*, or *The Systems of Nature*. Fresh out of university, Linnaeus used tables and visual diagrams to display the classification of the three kingdoms of nature: animal, vegetable and mineral. Due to their simplicity, these became popular very quickly and established Linnaeus' fame and reputation.

By the second edition in 1740, the tables were gone and genera of plants, animals and minerals were enumerated into an ever-expanding list, constantly added to and amended until the 13th edition, well after Linnaeus' death. (Plants were placed in their own book, *Species Plantarum*, in 1753.) The ground-breaking 10th edition of *Systema Naturae* in 1758 was the first zoological work to use Linnaeus' overarching legacy: binomial nomenclature (e.g. *Homo sapiens*), as well as taxa names such as Mammalia and Primates, which are still used today. **IC**

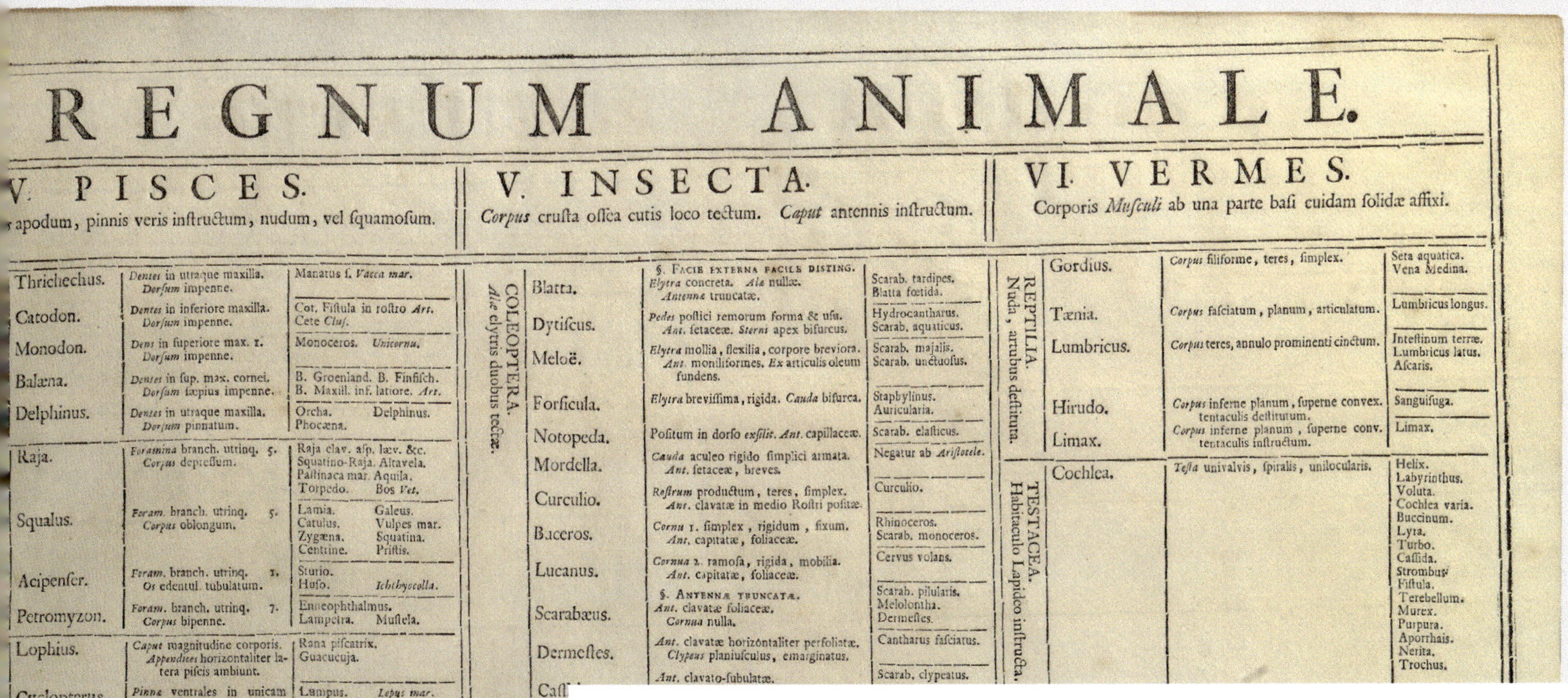

REGNUM ANIMALE.

V. PISCES.
…apodum, pinnis veris instructum, nudum, vel squamosum.

Trichechus.	*Dentes* in utraque maxilla. *Dorsum* impenne.	Manatus f. *Vacca mar.*
Catodon.	*Dentes* in inferiore maxilla. *Dorsum* impenne.	Cet. Fistula in rostro *Art.* Cete *Clus.*
Monodon.	*Dens* in superiore max. 1. *Dorsum* impenne.	Monoceros. *Unicornu.*
Balæna.	*Dentes* in sup. max. cornei. *Dorsum* sæpius impenne.	B. Groenland. B. Finfisch. B. Maxill. inf. latiore. *Art.*
Delphinus.	*Dentes* in utraque maxilla. *Dorsum* pinnatum.	Orcha. Delphinus. Phocæna.
Raja.	*Foramina* branch. utrinq. 5. *Corpus* depressum.	Raja clav. asp. læv. &c. Squatino-Raja. Altavela. Pastinaca mar. Aquila. Torpedo. Bos *Vet.*
Squalus.	*Foram.* branch. utrinq. 5. *Corpus* oblongum.	Lamia. Galeus. Catulus. Vulpes mar. Zygæna. Squatina. Centrine. Pristis.
Acipenser.	*Foram.* branch. utrinq. 1. *Os* edentul. tubulatum.	Sturio. Huso. *Ichthyocolla.*
Petromyzon.	*Foram.* branch. utrinq. 7. *Corpus* bipenne.	Enneophthalmus. Lampetra. Mustela.
Lophius.	*Caput* magnitudine corporis. *Appendices* horizontaliter lactera piscis ambiunt.	Rana piscatrix. Guacucuja.
Cyclopterus.	*Pinnæ* ventrales in unicam circularem concretæ.	Lumpus. *Lepus mar.*
Ostracion.	*Pinnæ* ventrales nullæ. *Cutis* dura, sæpe aculeata.	Orbis div. sp. Pisc. triangul. Atinga. Hystrix. Ostracion. Lagocephalus.
Balistes.	*Dentes* contigui maximi. *Aculei* aliquot robusti in dorso.	Guaperua. Histrix. Capriscus. *Caper.*
Gasterosteus.	*Membr. branch.* ossiculis 3. *Venter* laminis osseis instr.	Aculeatus. Spinachia. Pungitius.
Zeus.	*Corpus* compressum. *Squamæ* subasperæ.	Aper. Faber. *Gallus mar.*
Cottus.	*Membrana branch.* ossic. 6. *Caput* aculeatum, corpore latius.	Cataphractus. Scorpio mar. Cottus. *Gobio fl. capit.*
Trigla.	*Appendices* ad pinn. pect. articulatæ 2 vel 3.	Lyra. Gurnardus. Cuculus. Lucerna. Hirundo. Milvus. Mullus barb. & imberb.
Trachinus.	*Opercula branch.* aculeata. *Oculi* vicini in vertice.	Draco. Araneus mar. Uranoscopus.
Perca.	*Memb. branch.* ossicul. 7. *Pinnæ* dorsales. 1 vel 2.	Perca. Lucioperca. Cernua. Schraitser.
Sparus.	*Opercula branch.* squamosa. *Labia* dentes tegunt. *Dentes* molares obtinet.	Salpa. Melanurus. Sparus. Sargus. Chromis. Mormyrus. Mæna. Smaris. Boops. Dentex. Erythrinus. Pagrus. Aurata. Cantharus.
Labrus.	*Labia* crassa dentes teg. *Color* speciosus.	Julis. Sachettus. Turdus diversar. specier.
Mugil.	*Memb. branch.* ossic. 6. *Caput* totum squamosum.	Mugil. *Cephalus.*
Scomber.	*Memb. branch.* ossic. 7. *Pinnæ dorsi* 2 vel plures.	Glaucus. Amia. Scomber. Thynnus. Trachurus. *Saurus.*
Xiphias.	*Rostrum* apice ensiformi. *Pinnæ ventrales* nullæ.	Gladius.
Gobius.	*Pinnæ vent.* in 1 simpl. concr. *Squamæ* asperæ.	Gob. niger. Jozo. Paganellus. *Aphua.*
Gymnotus.	*Membr. branch.* ossicul. 5. *Pinna* dorsalis nulla.	Carapo.
Muræna.	*Membr. branch* ossic. 10. *Tubuli* in apice rostri. 2.	Anguilla. Conger. Fluta. Serpens mar.
Blennus.	*Pinnæ ventr.* constant oss. 2. *Caput* admodum declive.	Alauda non crist. & galer. Blennus. Gattorugine.
Gadus.	*Membr. branch.* ossic. 7. *Pinnæ dorsi.* 2 vel 3.	Asellus diversar. specier. Merluccius. Anthias 2dus. Mustela. Egresinus.
Pleuronectes.	*Membr. branch.* oss. 6. *Oculi* ambo in eodem latere.	Rhombus divers. specier. Passer. Limanda. Hippoglossus. Buglossus. Solea.

V. INSECTA.
Corpus crusta ossea cutis loco tectum. *Caput* antennis instructum.

COLEOPTERA. *Alæ* elytris duobus tectæ.	Blatta.	§. Facie externa facile disting. *Elytra* concreta. *Alæ* nullæ. *Antennæ* truncatæ.	Scarab. tardipes. Blatta fœtida.
	Dytiscus.	*Pedes* postici remorum forma & usu. *Ant.* setaceæ. *Sterni* apex bifurcus.	Hydrocantharus. Scarab. aquaticus.
	Meloë.	*Elytra* mollia, flexilia, corpore breviora. *Ant.* moniliformes. *Ex* articulis oleum fundens.	Scarab. majalis. Scarab. unctuosus.
	Forficula.	*Elytra* brevissima, rigida. *Cauda* bifurca.	Staphylinus. Auricularia.
	Notopeda.	Positum in dorso exsilit. *Ant.* capillaceæ.	Scarab. elasticus.
	Mordella.	*Cauda* aculeo rigido simplici armata. *Ant.* setaceæ, breves.	Negatur ab *Aristotele.*
	Curculio.	*Rostrum* productum, teres, simplex. *Ant.* clavatæ in medio Rostri positæ.	Curculio.
	Buceros.	*Cornu* 1. simplex, rigidum, fixum. *Ant.* capitatæ, foliaceæ.	Rhinoceros. Scarab. monoceros.
	Lucanus.	*Cornua* 2. ramosa, rigida, mobilia. *Ant.* capitatæ, foliaceæ.	Cervus volans.
	Scarabæus.	§. Antennæ truncatæ. *Ant.* clavatæ foliaceæ. *Cornua* nulla.	Scarab. pilularis. Melolontha. Dermestes.
	Dermestes.	*Ant.* clavatæ horizontaliter perfoliatæ. *Clypeus* planiusculus, emarginatus.	Cantharus fasciatus.
	Cass…	*Ant.* clavato-subulatæ.	Scarab. clypeatus.
	Chry…		
	Coc…		
	Gyr…		
	Nec…		
	Att…		
	Can…		
	Car…		
	Cici…		
	Lep…		
	Cer…		
	Bup…		
ANGIOPTERA. *Alæ* omnibus datæ, elytris destitutæ.	Pap…		
	Lib…		
	Eph…		
	Her…		
	Pan…		
	Rap…		
	Apı…		
	Ichn…		
	Mu…		
HEMIPTERA. *Alæ* elytris …	Gry…		Mantis.
	Lampyris.	*Pedes* 6. *Clypeus* planus. *Alæ* 4.	Cicindela.

VI. VERMES.
Corporis *Musculi* ab una parte basi cuidam solidæ affixi.

REPTILIA. Nuda, artubus destituta.	Gordius.	*Corpus* filiforme, teres, simplex.	Seta aquatica. Vena Medina.
	Tænia.	*Corpus* fasciatum, planum, articulatum.	Lumbricus longus.
	Lumbricus.	*Corpus* teres, annulo prominenti cinctum.	Intestinum terræ. Lumbricus latus. Ascaris.
	Hirudo.	*Corpus* inferne planum, superne convex. tentaculis destitutum.	Sanguisuga.
	Limax.	*Corpus* inferne planum, superne conv. tentaculis instructum.	Limax.
TESTACEA. Habitaculo Lapideo instructa.	Cochlea.	*Testa* univalvis, spiralis, unilocularis.	Helix. Labyrinthus. Voluta. Cochlea varia. Buccinum. Lyra. Turbo. Cassida. Strombus. Fistula. Terebellum. Murex. Purpura. Aporrhais. Nerita. Trochus.
ZOOP… Artub…			Penna marina.

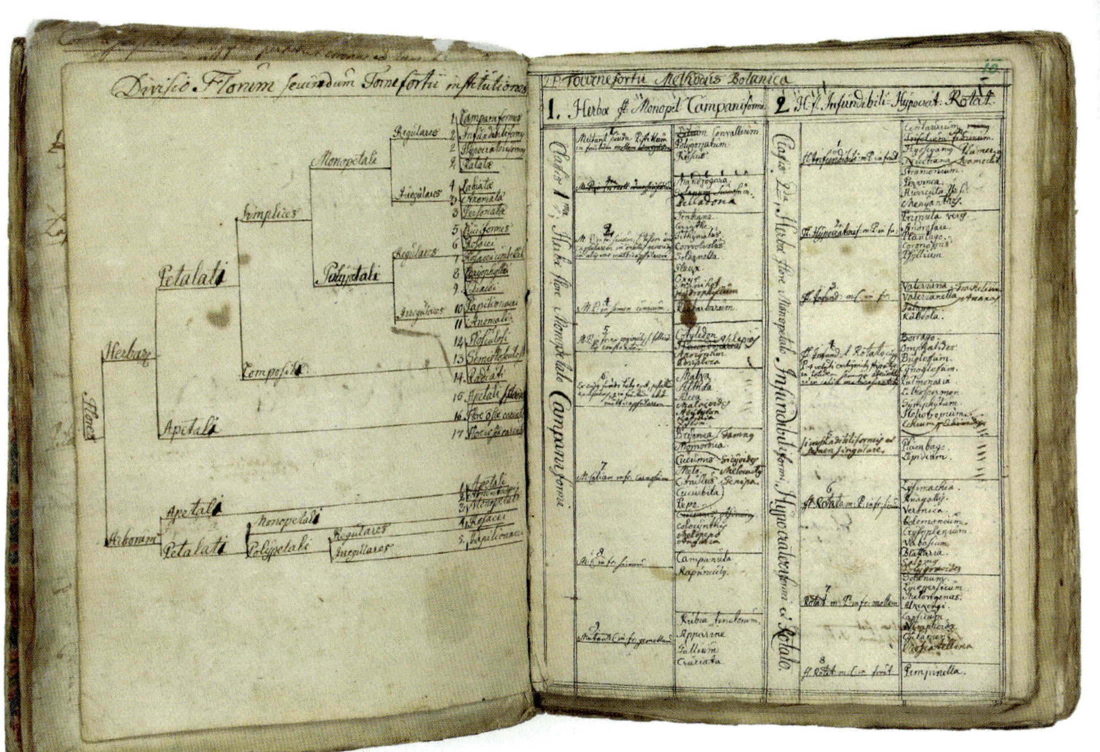

E Bibliotheca Linnæana propria.
1784. J E Smith.

HORTUS CLIFFORTIANUS

Plantas exhibens

QUAS

In Hortis tam VIVIS quam SICCIS,
HARTECAMPI in Hollandia,

COLUIT

VIR NOBILISSIMUS & GENEROSISSIMUS

GEORGIUS CLIFFORD

JURIS UTRIUSQUE DOCTOR,

Reductis Varietatibus ad Species,
Speciebus ad Genera,
Generibus ad Classes,

Adjectis Locis Plantarum natalibus
Differentiisque Specierum.

Cum *TABULIS ÆNEIS.*

AUCTORE

CAROLO LINNÆO,

Med. Doct. & Ac. Imp. N. C. Soc.

AMSTELÆDAMI. 1737.

Musa Cliffortiana & *Hortus Cliffortianus*

1735

In the autumn of 1735, Carl Linnaeus was invited to become garden superintendent and personal physician to George Clifford (1685–1760), a very wealthy Anglo-Dutchman and director of the Dutch East India Company. At the time, bananas were an almost unknown curiosity in northern Europe and cultivating them to bear fruit was proving difficult; even the respected French botanist Antoine de Jussieu (1686–1758) had not managed it. One of the first tasks that Linnaeus set himself was to produce flowers and fruit on Clifford's plant at his Hartecamp estate in North Holland. Linnaeus and the gardener, Dietrich Nietzel, changed the heating and watering regime of the plant, and in 1736 they were successful; Linnaeus even sent a banana to Jussieu. This was allegedly the first time that a banana had flowered and fruited so far north.

The achievement was met with fascination, and many came to gaze on the fabulous plant. Linnaeus capitalised on the opportunity by quickly publishing *Musa Cliffortiana* (1736) which celebrated his achievement and gave him an opportunity to further expand on his 'sexual system' of plant classification and discuss its merits in relation to 'natural' systems. The work was beautifully illustrated by Martin Hoffman and depicted the whole plant, flowers and developing fruit (Right).

The culmination of Linnaeus' stay at Hartecamp was the production of *Hortus Cliffortianus* (1737—Left), illustrated with engravings by Jan Wandelaar based on the work of Georg Dionysius Ehret, the greatest botanical artist of the age. The frontispiece is lavishly allegorical: to the left of Cybele (Mother Earth) strides Apollo, with the face of a youthful Linnaeus. If that were not enough, towering above Apollo's head, a fruitful banana tree sways in the breeze. Shown here is Linnaeus' own copy. **MAS**

Portrait of a Professor

1736

This portrait, set in a gilded oval frame, sits in contrast with the heavy, dark portraits of so many of the Society's eminent members. It shows the contented smile and stout figure of the first Sherardian Professor of Botany at Oxford Botanic Garden, Johann Jakob Dillenius (1687–1747). Presented to the Society by W. Carruthers on 24 May 1889, this is a copy of the original portrait held at the University of Oxford (artist and date unknown).

Dillenius, a German botanist, was invited to come to England by fellow botanist William Sherard (under whose name the Oxford professorship was awarded), which he did in August 1721. Dillenius eventually ended up working for William's brother James, a successful apothecary and keen cultivator of exotic plants in Eltham, Kent. The result was the remarkable *Hortus Elthamensis* (1732), probably the first garden catalogue, for which Dillenius not only wrote the text, carrying out detailed botanical and nomenclatural research, but also completed the 324 illustrations. One of the plants illustrated was *Sprekelia formosissima* (OPPOSITE), known at the time as *Lilionarcissus jacobaeus flore sanguineo nutante*, which Dillenius is also holding in the portrait.

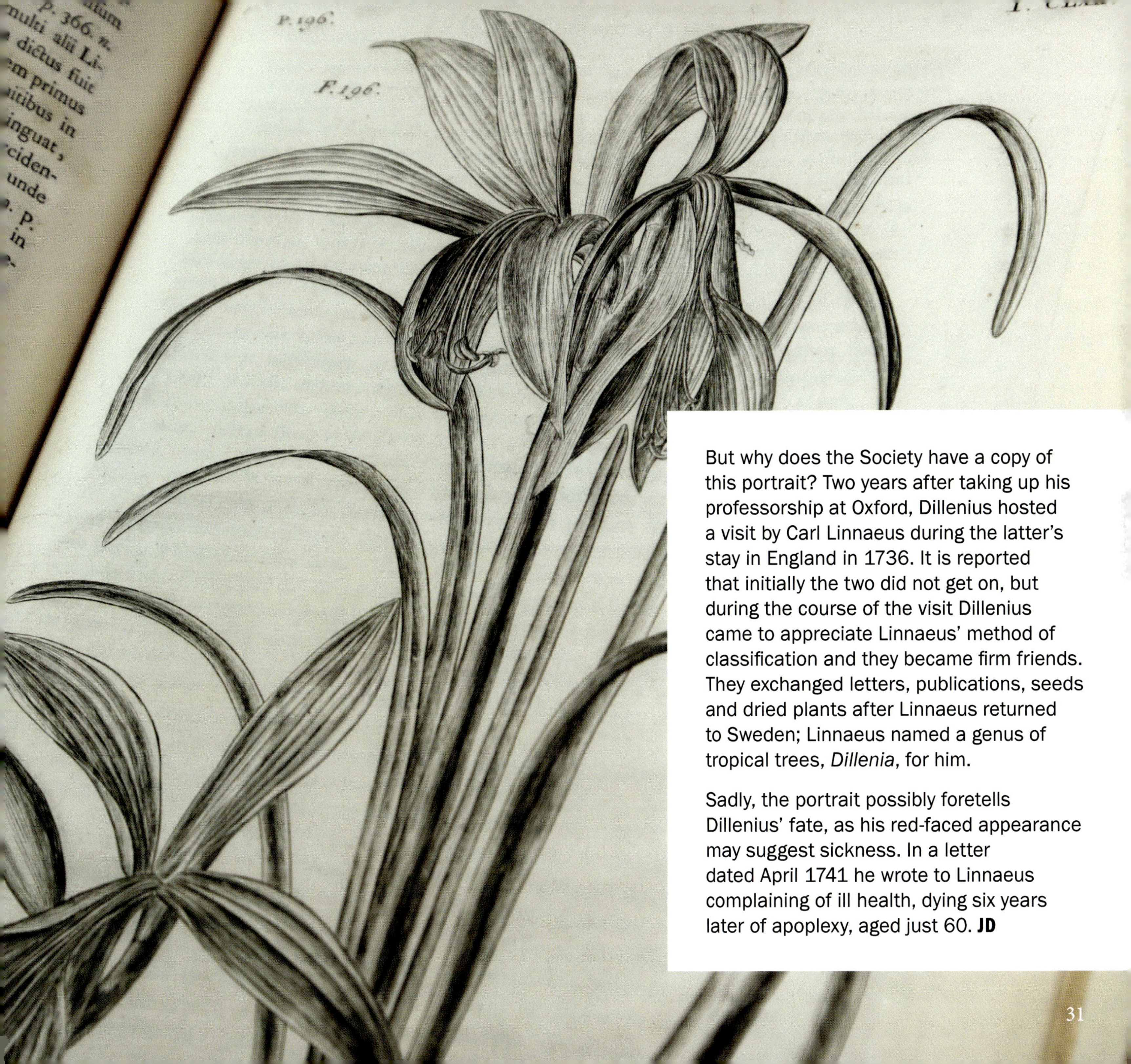

But why does the Society have a copy of this portrait? Two years after taking up his professorship at Oxford, Dillenius hosted a visit by Carl Linnaeus during the latter's stay in England in 1736. It is reported that initially the two did not get on, but during the course of the visit Dillenius came to appreciate Linnaeus' method of classification and they became firm friends. They exchanged letters, publications, seeds and dried plants after Linnaeus returned to Sweden; Linnaeus named a genus of tropical trees, *Dillenia*, for him.

Sadly, the portrait possibly foretells Dillenius' fate, as his red-faced appearance may suggest sickness. In a letter dated April 1741 he wrote to Linnaeus complaining of ill health, dying six years later of apoplexy, aged just 60. **JD**

211. BUPRESTIS
Buprestis
2. 8-guttata
1. gigantea

Linnaeus' Insects

1752

In 1752, Carl Linnaeus was asked to catalogue the Swedish royal natural history collections. Amongst Queen Lovisa Ulrika's insect collection were specimens of a large 'Metallic Wood Boring Beetle' that lives in the warm elevated regions of Central and South America. Linnaeus named it *Buprestis gigantea* (now *Euchroma gigantea*). To describe it in his various editions of *Systema Naturae*, Linnaeus relied on these specimens as well as previous literature, including two famous 17th-century works: Hans Sloane's *A Voyage to ... Jamaica* (1725) and Maria Sybilla Merian's stunning *Metamorphosis Insecta Surinamensis* (1705), where the beetle is pictured with its larva on the white potato plant upon which it feeds (Left).

The Linnean Society possesses four specimens of *Buprestis gigantea* (Boxed—Left), their beautiful iridescence unaltered by age. Because the Society's founder, James Edward Smith, added his own specimens to the Linnaean entomology collection, it is often hard to identify each specimen's provenance. However, the specimen furthest to the left, wings closed, is clearly Linnaean in origin (Linnaeus' label for 'gigantea' is obscured by the insect here). It is possible that this specimen originally came from the Queen's collection, though it might have been sent by the Spanish priest and botanist José Celestino Mutis, who sent numerous letters, specimens and illustrations to Linnaeus from Central America.

Other specimens were later added to the collection by Smith: one from Cayenne in French Guiana made its way to Smith via a W. Roe from Norfolk; another specimen, with its wings outstretched, was sent by the German merchant Carspar von Voght from Surinam when he corresponded briefly with Smith in the late 1790s.

The four specimens of *Buprestis* are characteristic of the composite quality of the Society's insect collections, as well as of specimen exchange as networking currency in the late 18th and early 19th centuries. **IC**

LSL INS 873

What's in a Name?

1753

The herbarium of Carl Linnaeus is one of the most important natural history collections in the world. Its value cuts across many aspects of botanical science, but of paramount importance is the significant quantity of type specimens—the physical object upon which a scientific name is based. Many of these type specimens are of important agricultural crops such as rice, potato and carrot. Others are ecologically significant wild plants, or those found in gardens.

This specimen is of a common lime (*Tilia × europaea*), a native European species that can be found in many gardens and parks. It is a naturally occurring hybrid between the small-leaved lime (*T. cordata*) and the large-leaved lime (*T. platyphyllos*); Linnaeus allegedly collected this specimen from the vicinity of Uppsala Cathedral in his native Sweden, prior to the publication of his botanical opus, *Species Plantarum* (1753).

Lime trees probably held an important place in Linnaeus' heart, as his family name originated from the Swedish for lime: *lind*. In the late 17th century, near the family farm of Jonsboda in Småland, there was a fine three-trunked specimen of small-leaved lime. It is believed that this inspired Carl's father Nils to adopt 'Linnaeus' as the family name. Relatives of Linnaeus also named themselves after the same tree, although they called themselves Tiliander, after the lime tree genus *Tilia*. The origins of the name *Tilia* are disputed but it is probable that the name originates from the Greek word *tilos*, meaning thread or fibre, for which lime trees had long been used. **MAS**

Fumaria
881.1
117–136
137–183
428–458
499–522
218–241

Order out of Chaos

1753

Preserving plant specimens by pressing and drying them for subsequent study was a 16th-century invention in Italy. Carl Linnaeus recognised the scientific value of such collections and his own herbarium of pressed plants at the Linnean Society contains more than 14,000 sheets.

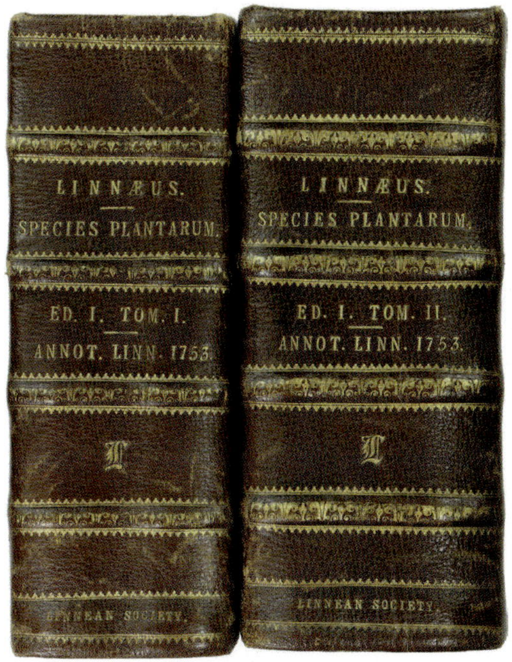

One of Linnaeus' innovations was to introduce a new, two-name (binomial) system in Latin for the naming of organisms: each species has a genus name (which it may share with other species) to which is added a specific, or species name. On this pressed specimen of 'Dutchman's breeches' (Opposite Left), a member of the poppy family, Linnaeus wrote the genus as 'Fumaria' and species as 'Cucullaria'. *Fumaria cucullaria* had been known by the altogether less succinct '*Fumaria siliquosa, radice grumosa, flore bicorporeo ad labia conjuncto, virginiana*'. Linnaeus' naming system provided a unique, more memorable binomial name for any given plant.

It was in his ground-breaking botanical work *Species Plantarum* (1753) that Linnaeus introduced binomial nomenclature to the world (Right). He arranged the species in the sequence of his 'sexual system'; a method of grouping plants based on the number and arrangement of the male (stamens) and female (pistil) parts in the flower. Within this system he created 24 'Classes' based on the number and arrangement of the stamens, each subdivided into 'Orders' grouped on other reproductive features.

To house his pressed specimens, Linnaeus designed tall herbarium cabinets with shelves that followed the sequence of his 'sexual system' of classification (Left). Previously, plant specimens were mounted in bound volumes (making it difficult to add to, or rearrange, them), so Linnaeus pasted each plant on a separate sheet, allowing for easy rearrangement and addition of new material to the shelves. And though this 'sexual system' had fallen out of favour by the early 19th century, the way in which we house plant specimens and name species is broadly the same today. **CJ**

Alexander Garden's Fish Specimens

1761

Nowadays, scientific specimens of Fishes are generally preserved whole in alcohol. But in the 18th century this technique was comparatively new and most collectors followed a method established by Johan Frederic Gronovius (1730–1777), where fish were mounted rather like herbarium specimens on sheets of paper. Carl Linnaeus certainly did possess many alcohol-preserved specimens, but only his mounted specimens were sent with the collections to London, due to the relative ease in transporting them.

There are 168 such dried fish specimens in the Linnean Society's collections, which provided sufficient information for initial descriptions of species, such as the overall body shape, number of rays in the fins, number of scales in a lateral series, etc. A number have been identified as type specimens, used in Linnaeus' original descriptions of the species.

Three contributory sources are discernible in this fish collection, the largest being 87 specimens sent to Linnaeus from South Carolina by Scottish naturalist Alexander Garden (1730–1791) in 1761. A further 49 form what has been called the Scandinavian Series, many of which were possibly collected by Linnaeus himself, and the remaining 32 specimens may have been collected by the German naturalist Nathanael Gottfried Leske (1751–1786).

Shown here with its likeness in Marcus Bloch's *Ichtyologie ou histoire naturelle générale et particulière des poissons* (1785–1797), these specimens of remora or sharksucker (*Echeneis naucrates*) were sent to Linnaeus by Alexander Garden and are among the earliest surviving specimens from North America. Garden's characteristic paper label around the tail of one of the fish bears his inscription: 'Echeneis nostratib, Sucking Fish.'

The specimens illustrate how the fish can become distorted if they are not laid out flat on the backing paper before the **drying process. However, clearly visible on the top of each head is the prominent** sucking disc, shaped like the sole of a shoe, which this species uses to attach itself to sharks, turtles and even boats. **OC**

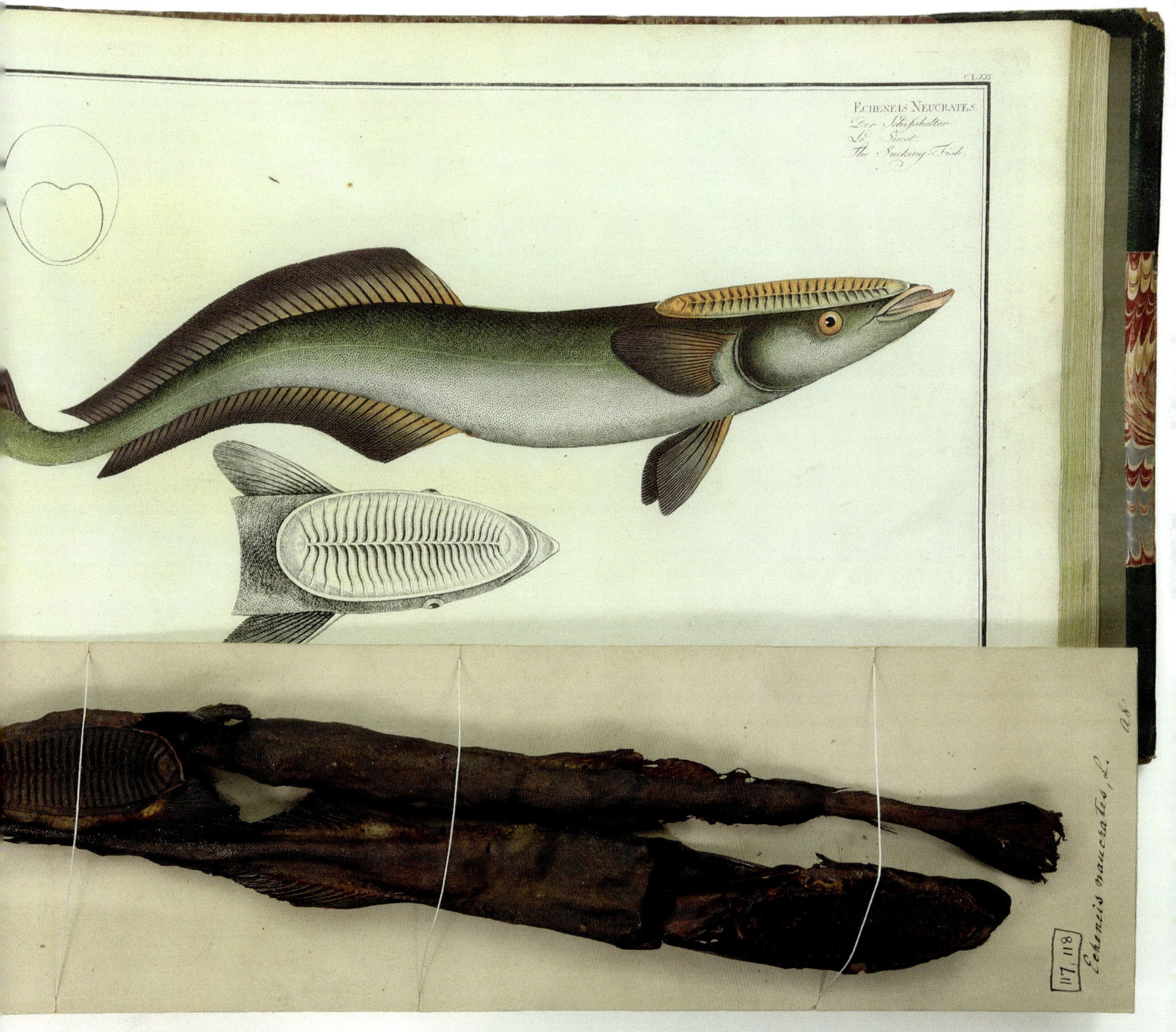
CLXXI
Echeneis Neucrates.
Der Schiffshalter.
Le Sucet.
The Sucking-Fish.
Echeneis naucrates, L.
117.118

Pearls of Wisdom

1761

Carl Linnaeus once said that he had 'heard of people who could make gold, but had never heard of anyone who could make pearls'. Linnaeus had journeyed to Lapland in 1732 where he saw natural pearls in the village of Purkijaur. Throughout his life, Linnaeus was interested in the connection between natural resources and their economic value, and by 1761 he believed that there was a way to boost Sweden's wealth through pearl culturing.

His tests were based on an old Chinese method for culturing blister pearls (which are flat on one side); after drilling a hole in a mussel's shell, he inserted a granule of limestone between the shell and the mantle. (This was based on the 'myth' that pearls are produced from a grain of sand, whereas it is now understood that they are a response to infection.) The mussels were then returned to the riverbed for several years, and produced what are believed to be the first artificially cultured spherical pearls.

Linnaeus wrote of his success to his friend Carl Hårleman, and the commercial potential of the project was discussed in both the Swedish Chamber of Commerce and the Swedish Parliament. Linnaeus was prepared to reveal his method in return for the right to nominate his successor as Professor of Medicine at Uppsala University. There was also talk of awarding Linnaeus the not insignificant sum of 12,000 riksdaler for his work, though in the end half of that sum was settled upon.

The patent for cultured pearls was sold to a merchant named Peter Bagge, but ultimately nothing came of the venture. While the idea did not increase the national wealth, Linnaeus was satisfied; he used the money awarded to him to pay off the debt on his small estate of Hammarby, just outside of Uppsala, and was ennobled 'Carl von Linné' by the King of Sweden for this and other efforts. **GBr**

A Singular Plant: *Roussea simplex*

1771

Jean-Jacques Rousseau (1712–1778), the 18th-century French philosopher, remains as controversial today as he was in his own lifetime. He has been variously praised or blamed for begetting nationalism, Romanticism, collectivism and the French Revolution. As a nonconformist, James Edward Smith felt a particular affinity with Rousseau's advocacy for natural religion and rejection of Church hierarchy. Even after the commencement of the Terror in France made Rousseau's name anathema in polite English society, Smith defended the ideals of his hero.

With his purchase of the Linnaean collections in 1784, Smith came into possession of a letter from Rousseau to Carl Linnaeus. In later life, Rousseau became enamoured of botany, and, on 21 September 1771, wrote a fawning letter to the great Swedish naturalist: 'I read you, I study you, I meditate upon you, I honour you and love you with all my heart. J.J. Rousseau.'

During his Grand Tour of 1786–1787, Smith undertook a pilgrimage to Ermenonville, near Paris, Rousseau's first burial place. Smith saw Rousseau's snuffbox and the shoes he died in, preserved by the local landlord. At Rousseau's tomb Smith reverently collected several fragments of the lichen *Umbilicaria murina*. He also visited Marie-Thérèse Levasseur, Rousseau's widow, who showed him Rousseau's death mask. It had, Smith wrote, a 'great serenity in the countenance, & much sensibility. The mouth is uncommonly beautiful'.

In 1789 Smith fulfilled Linnaeus' intention to name a genus after Rousseau. He chose an unnamed specimen from the herbarium of Linnaeus' son, a curious plant collected from the island of Mauritius by Philibert Commerson. *Roussea simplex* is a singular plant, pollinated by geckos (a fact unknown to Smith), and remains the only species of its genus. **TK**

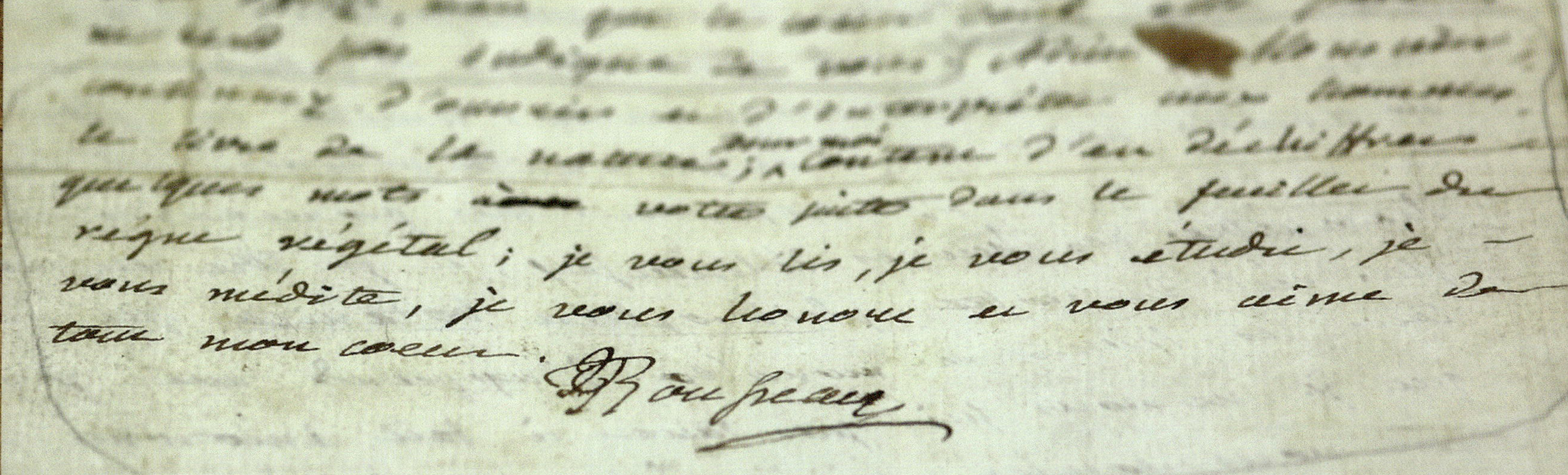
quelques mots votre dans le feuillet du règne végétal; je vous lis, je vous étudie, je vous médite, je vous honore et vous aime de tout mon coeur.
J.J. Rousseau

of the Paffage from
ENGLAND
to
BOTANY BAY
in
EWHOLLAND.
1787.
OPE
I C A
Norfolk Island
Common Blue Vine
Large White Vine
Norfolk Island
Flax Plant
Norfolk Island
Cabbage Palm
Cancer
CONGO
HOTTENTOTS
Straits
King William Cape
Guinea
Endeavour R.
NEW HOLLAND.
New South Wales
Land of Peter Nuyts
I. St. Francois
I. St. Pierre
Land of Lions
South Cape
Diemens Land discoverd 1642.
Prince Edward Isles
from
to
Botan

Seeds of the South Pacific

1774

On 10 October 1774, during Captain James Cook's second voyage, those on board the HMS *Resolution* caught sight of what became known as Norfolk Island.

The seed packets shown here, complete with rough sketches of the relevant plants, are a bit of a mystery: were they brought back by Georg Forster, who was, with his father Johan Reinhold Forster, a naturalist on board the *Resolution*, or were they collected later by Philip Gidley King, who led a party of 15 convicts and seven free men from the First Fleet onto the island in 1788?

The handwriting is close to that of Georg Forster, who observed the New Zealand flax and cabbage palm on Norfolk Island. Forster named the New Zealand flax *Phormium tenax* in 1776 and described both plants in his *Voyage Round the World* (1777). The sketches and use of vernacular names seem to indicate that these seeds were more likely to have been collected and labelled in October 1774 rather than later, once a Latin name had been given.

Whoever the collector was, the seed packets made their way to Linnean Society founder James Edward Smith, via the botanist Aylmer Bourke Lambert (1761–1842).

The seeds collected held economic interest. The growing tip of the cabbage palm (the only native palm), also known as the niau or Norfolk Island palm (*Rhopalostylis baueri*), is an edible food source. Additionally, the British hoped to develop an alternative flax industry to compete with that of Russia, as flax was essential in the making of sails and rope for the growing British naval fleet.

However, these plans failed, as they were unaware of the techniques necessary to 'dress' or process flax. In 1793, Philip Gidley King instructed that two Northland Māori Chiefs be kidnapped to demonstrate the necessary skills—only to be told these practices were usually undertaken by women. **MAS, IC, MJ**

Signed, Sealed, Delivered

1784(–1932)

When Sir James Edward Smith acquired the Linnaean collections in 1784, he set about building a collection of seals, using as a starting point the rich correspondence that Carl Linnaeus enjoyed with fellow Swedes and foreign naturalists from 1728 until his death in 1778.

The Linnaean Correspondence Project estimates that about 5,500 letters were exchanged between Linnaeus and his more than 600 correspondents. Of these, over 3,300 are preserved at the Linnean Society, having also been scanned to view online. Smith carefully selected and cut out what he deemed to be important seals and displayed them in a beautiful mahogany cabinet.

Four of the 13 seals used by Linnaeus in his lifetime are present, including a very rare one (below, furthest to the right). Also represented are seals of famous naturalists, such as Joseph Banks, William Aiton, Antoine Laurent and Bernard de Jussieu, and Albrecht von Haller.

Smith enriched the collection by adding seals from his own network. After Smith's death in 1828, the cabinet was left to his descendants. His great grand nephew, Francis Pierrepoint Barnard, departed from Smith's original project by adding more antiqurian seals: antique heads, societies seals, etc. At Barnard's death in 1932, his widow offered the cabinet to the Linnean Society, and it is now possible to reunite, at least virtually, some seals to their original letters. **IC & EN**

William Markwick's Nature Calendar

1789–1800

William Markwick (1739–1813) was an English country gentleman who lived at Catsfield, near Battle, in the south of England, some five miles from the Sussex coast. Markwick was a friend and correspondent of the better-known Gilbert White of Selbourne (1720–1793), and both men kept 'nature calendars', recording the dates on which they first and last saw such migrant birds as the barn swallow (*Hirundo rustica*), European nightjar (*Caprimulgus europaeus*) and European wryneck (*Jynx torquilla*). Studies of 'phenology', as the timing of natural events became known, started with gardening calendars in the early 1700s, and those of Markwick and White have been invaluable in understanding the phenological changes associated with climate change.

Markwick, a keen naturalist, was elected to the Linnean Society in 1792. He contributed a number of papers, some of which were published in the Society's *Transactions*, including one on the migration of birds in 1789. A decade later he published *Aves Sussexiensis or, a Catalogue of Birds Found in the County of Sussex* (1798), and in April 1800 he presented the Society with his illustrated manuscript entitled 'Remarks on Birds'.

The table of 'first seen and last seen' migratory birds in his 1789 article makes interesting reading, partly because late sightings, such as those of swallows (*Hirundo rustica*) seen in mid-November, reinforced the (then still prevalent) view, that these birds might hibernate over winter rather than migrate. Markwick's table notes that between 1768 and 1789, the first swallows of spring were not seen earlier than 7 April; prior to this the average first sighting of the birds was 14 April. Climate change has brought forward the arrival dates of many migratory birds, yet even though Britain was enduring the Little Ice Age during the period Markwick's records were made, his dates are surprisingly similar to those of recent times. **TRB**

A Table of the appearance & disappearance of
The Swallow
Hirundo rustica, first seen
not seen after
The Martin
Hirundo urbica first seen
not seen after
The Swift
Hirundo Apus first seen
not seen after
The Sand Martin
Hirundo riparia first seen
not seen after
The Wry Neck
Jynx torquilla
The Redstart
Motacilla phœnicurus
The Willow Wren
Motacilla trochilus
not seen after
The Nightingale
Motacilla luscinia first seen
not seen after
The Swallow. Hirundo rustica
Synonyma
The Chimney Swallow. Lath. Syn. IV. p. 561. n. 1
Hirundo rustica
H. nigro-cœrulescens subtus albida, fronte
gulaque castaneis, rectricibus lateralibus
macula alba notatis. Lath. Orn. II. p. 572. n. 1.
Hirundo rustica
H. rectricibus, exceptis duabus intermediis,
macula alba notatis. Linn. Syst. p. 343. n. 1.
The House Swallow. Britt. Zool. Fol. Ed. p. 96
Description
Its Length from the Tip of the Bill to the End
of one of the longest Feathers of the Tail is
seven Inches and an half; its Breadth from
Tip of the Wing to Tip of the Wing, when extended,
thirteen Inches and an half.
Its Bill is small, short, slender, and of a
black Colour; its Mouth very wide and of a
pale yellow withinside; its Forehead and
Throat are of a dull reddish Colour; its

J Tyley del
Artocarpus incisus

Alexander Anderson & John Tyley's Caribbean Drawings

1794

Scottish naturalist Alexander Anderson (1748–1811) and artist John Tyley (fl. 1794-1823) lived through turbulent times.

Having emigrated to New York in 1774, Anderson was briefly involved in the American War of Independence, captured by privateers, and imprisoned by French forces. He also experienced an attempted revolution in the Caribbean. Anderson served as the second director of the botanical garden on the island of St Vincent from 1785 until a few months before his death. It flourished under his administration, as Anderson brought more than 1,300 new species of plants, many of them exotic, into the garden.

Tyley was described by Anderson in a letter to botanist William Forsyth as 'a deserving young man, a (...) self taught (...) Mulatto native of Antigua'. By 1794, Anderson was employing Tyley as a botanical artist to illustrate a planned catalogue of the plants in the botanical garden. However, Tyley's political activities led to his expulsion from St Vincent in 1800. He is last traced in Antigua in 1823, having signed a petition to improve the lives of people of colour.

Anderson's catalogue was never published, but three manuscripts drafts, as well as a corpus of 147 drawings of plants and a garden map, are held at the Linnean Society. Tyley signed 11 of these drawings, and it is likely that he also painted many of the unsigned ones.

In 2020, the Linnean Society acquired his extraordinary painting of *Artocarpus incisus*. Departing from the botanical drawings he made for Anderson, it depicts the breadfruit, a tree imported to St Vincent in 1793 to feed the enslaved population, some of whom worked in the garden and were owned by Anderson. What makes this painting remarkable is the presence of an enslaved man at the foot of the tree in a stance far from subservient. The triangular shape of the tree also recalls Revolutionary France's tree of liberty. **IC & JK**

Australia's First Fleet: *Eucalyptus robusta*

1795

In 1787, Dr John White (c. 1756–1832) set sail for Australia as the appointed surgeon on board the convict transport ship *Charlotte*, as part of the famous First Fleet. He later became Surgeon General to the newly formed colony of Port Jackson. Shortly after arriving, White became interested in the plants and animals of the area.

He investigated the medicinal properties of Australian plants and most likely gained some of his knowledge from the indigenous Eora people. The various aromatic mallee bushes, gulgong and mugga trees interested the medical staff of the fleet; both he and surgeon John Considen successfully distilled oil from the peppermint gum, *Eucalyptus piperita*, in 1788 (though it is disputed who did it first).

White sent numerous plant specimens to the Society's founder, James Edward Smith, who incorporated them into his collection where they remain today. Smith used these collections when describing Australian plants in his serial work *A Specimen of the Botany of New Holland* (Right) between 1793–1795. This work is considered to be the first publication solely dedicated to the flora of Australia, and was enriched by the illustrations of James Sowerby. The gulgong tree of the Eora was described by Smith in 1795 as *Eucalyptus robusta*, and was illustrated by Sowerby using the specimens collected by White in Port Jackson in 1793 (Opposite Right). The potential value of this tree as a timber crop was already recognised by Smith who provided the alternative English name of New Holland Mahogany. **MAS**

39
EUCALYPTUS robusta.
Brown Gum Tree, or New Holland
ICOSANDRIA
GEN. CHAR. Cal. superus, persisten
anthesin tectus operculo integer
nulla. Caps. quadrilocularis,
sperma.
Cal. superior, permaner
flowering with an
none. Caps. of 4
ing many seeds.
SPEC. CHAR.
Port Jackson, New South Wales, Mr White 1793.
876.20
876

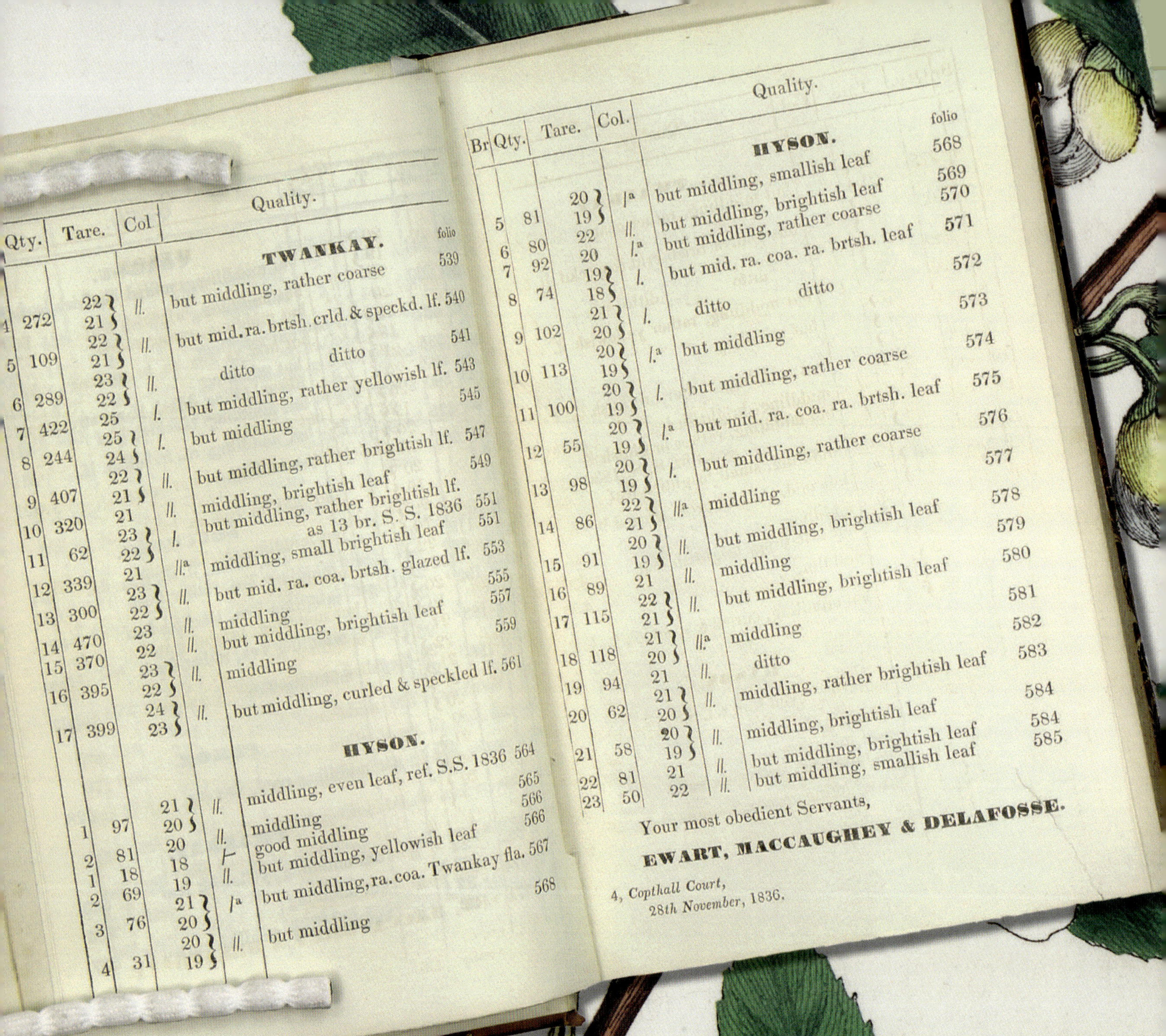

	Qty.	Tare.	Col	Quality.	folio
				TWANKAY.	
4	272	22, 21	II.	but middling, rather coarse	539
5	109	22, 21	II.	but mid. ra. brtsh. crld. & speckd. lf.	540
6	289	23, 22	II.	ditto ditto	541
7	422	25	I.	but middling, rather yellowish lf.	543
8	244	25, 24	I.	but middling	545
9	407	22, 21	II.	but middling, rather brightish lf.	547
10	320	21	II.	middling, brightish leaf	549
11	62	23, 22	I.	but middling, rather brightish lf. as 13 br. S. S. 1836	551
12	339	21	II.[a]	middling, small brightish leaf	551
13	300	23, 22	II.	but mid. ra. coa. brtsh. glazed lf.	553
14	470	23	II.	middling	555
15	370	22	II.	but middling, brightish leaf	557
16	395	23, 22	II.	middling	559
17	399	24, 23	II.	but middling, curled & speckled lf.	561
				HYSON.	
1	97	21, 20	II.	middling, even leaf, ref. S.S. 1836	564
2	81	20	II.	middling	565
1	18	18	⊢	good middling	566
2	69	19	II.	but middling, yellowish leaf	566
3	76	21, 20	I.[a]	but middling, ra. coa. Twankay fla.	567
4	31	20, 19	II.	but middling	568

Br	Qty.	Tare.	Col.	Quality.	folio
				HYSON.	
5	81	20, 19	I.[a]	but middling, smallish leaf	568
6	80	22	II.	but middling, brightish leaf	569
7	92	20	I.[a]	but middling, rather coarse	570
8	74	19, 18	I.	but mid. ra. coa. ra. brtsh. leaf	571
9	102	21, 20	I.	ditto ditto	572
10	113	20, 19	I.[a]	but middling	573
11	100	20, 19	I.	but middling, rather coarse	574
12	55	20, 19	I.[a]	but mid. ra. coa. ra. brtsh. leaf	575
13	98	20, 19	I.	but middling, rather coarse	576
14	86	22, 21	II.[a]	middling	577
15	91	20, 19	II.	but middling, brightish leaf	578
16	89	21	II.	middling	579
17	115	22, 21	II.	but middling, brightish leaf	580
18	118	21, 20	II.[a]	middling	581
19	94	21	II.	ditto	582
20	62	21, 20	II.	middling, rather brightish leaf	583
21	58	20, 19	II.	middling, brightish leaf	584
22	81	21	II.	but middling, brightish leaf	584
23	50	22	II.	but middling, smallish leaf	585

Your most obedient Servants,

EWART, MACCAUGHEY & DELAFOSSE.

4, Copthall Court,
28th November, 1836.

MARCH SALE, 1797.

TEA

FOR

SALE

AT THE

EAST INDIA-HOUSE,

To follow the Dutch Teas for Exportation,

On TUESDAY the 12th of JUNE, 1797.

With all Faults, in Time, to advance 1d.

Rare Tea Auction Catalogues

1797 & 1836

The Insch Tea Library is a treasure trove of tea history. James Insch FRMS FLS (1876–1951) was a tea planter and Senior Director of Messrs Duncan Brothers & Co. Calcutta, as well as Chairman of the Indian Tea Association. Tea arrived in England in 1660, soon after its introduction to Holland (1610) and France (1630). By 1662 Catherine of Braganza (wife of King Charles II) had introduced tea to high society in England, which ensured its popularity.

Tea auctions soon followed in 1679–1680, with four major auctions held per year under the control of the British East India Company (who had the sole rights to import tea from China). Initially they were called 'Candle Light Auctions', as tea was 'sold by the candle'; as each lot was announced, a candle marked in inches was lit, and the hammer fell on each lot as the markers were reached. Tea auctions were loud affairs: it was said that in spite of the thick walls of East India House on London's Leadenhall Street, the shouting could still be heard in nearby Leadenhall Market. By 1750 a staggering 4.72 million lbs of tea were imported into Great Britain. By 1797, English tea drinking hit a rate of 2 lbs per capita annually, a rate that increased by five times over the next 10 years.

Tea catalogues were published for each of these auctions. Very few still exist and the Linnean Society has two, from 1797 and 1836 (the year the British East India Company lost its monopoly on the trade). The catalogues listed details of the ship that transported the tea, date of inspection by Customs, the warehouse where it was stored (by code) and the number of chests. Every chest of tea was inspected, sampled by specially designated tea brokers (around 30 in all) and given its own quality mark in the catalogues. The page of one broker, Ewart, Maccaughey & Delafosse, is displayed here. Tea auctions continued until 1998—a period of 319 years. **RB**

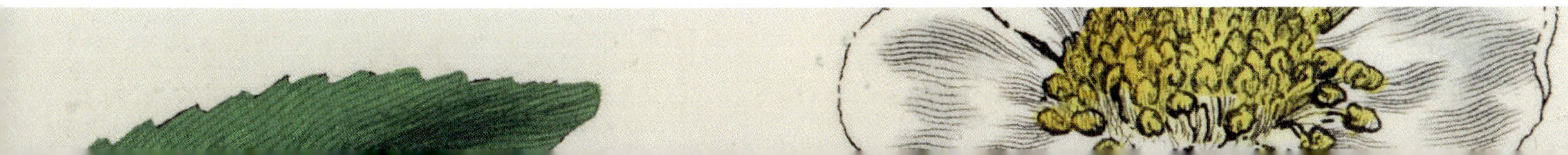

19th Century

The 19th century saw the birth of science as a profession, characterised by rapid technological changes and paradigm shifts like Charles Darwin and Alfred Russel Wallace's theories of evolution by natural selection, and Brownian motion. The specimens, papers and letters sent in by Fellows contributed to the enrichment of the Society's collections, and illustrate Britain's colonial past.

Richard Dreyer's Illustrated *Flora Britannica*

1800s

Some of the most beguiling objects in the Society's library collections are those which show traces of their former owners. A select number of items contain elaborate marginalia, including the wonderful copies of Society founder James Edward Smith's *Flora Britannica* (1800–1804), extensively annotated and illustrated by the Reverend Richard Dreyer.

Dreyer (1763–1838) studied at Trinity Hall, Cambridge, and became a Fellow of the Linnean Society in 1817, enjoying a lively correspondence with many eminent botanists and entomologists of the day.

His copies of Smith's *Flora* reveal eye-catching marginal illustrations. It has been asserted that Dreyer based his drawings not on field observations, but on plates from contemporary published works like James Sowerby's monumental *English Botany* (1790–1814).

Dryer drew up his will in 1818, pledging both the *Flora* and his collection of drawings and manuscripts to the Society. However, in a codicil of 1824, he withdrew the gift. The reasons for this change of heart are unclear, but it is possible a hardening in Dreyer's attitudes towards religious dissenters may have been a contributing factor. Smith came from a notable Unitarian family, and his application for the position of Chair of Botany at Cambridge University sparked a bitter debate. At a time of increasing division among both the Linnean Fellowship and wider society, perhaps Dreyer had chosen his side?

The full story is lost to history. In any case, Dreyer's widow donated the *Flora* in 1838. Its charismatic presentation, married to an expert eye for botanical detail, make it a firm favourite with both scholars and the general public. **WB**

POLYADE
POLYANDR

Linn.Trans.VI. tab. 22. p.207
T. Davies del. 1799.
SP 294
MENURA
No 238
Char. Gen.
Rostrum validiusculum
Nares ovatæ in medio
Pedes validi ambulato
Menura superba

Thomas Davies: A Soldier Naturalist

1800

On 4 November 1800, five papers by Major General Thomas Davies (c. 1737–1812) were read to an audience of 24 Fellows. The papers were varied and included accounts of the platypus, superb lyre bird and the narwhal (Below). As is indicated on the accompanying drawing, the narwhal was, in length, '24 feet 6, taken near Boston Lincolnshire March 1800'. The Society still preserves three of the original drawings for these papers, all of which were painted by Davies in the year 1800.

That Davies could submit descriptions and drawings of animals from places as disparate as Australia, Lincolnshire and Bengal is due to the fact that he was a well-travelled officer in the British Army. Davies began service in Canada in the late 1750s, participating in the French and Indian War, then later in the American Revolutionary War, before being assigned command posts in Gibraltar, the West Indies and Canada. Having been trained in military topographical drawing, he produced many landscapes, with particular attention to flora and fauna. During a six-year stint in England, the Society's archives show that Davies made connections with other naturalists like John Ellis, who invited the young captain to call on him in the spring of 1770. Davies' fondness for natural history inspired in him a desire to share details of 'the rarer animals which I have met with in those distant Countries where my Duty has called me'. He was elected as a Fellow of the Linnean Society in 1791, having been a Fellow of the Royal Society since 1781.

His account of the *Menura superba*, or superb lyre bird (Left), is the first to scientifically illustrate and describe the species (now *Menura novaehollandiae*), an Australian songbird famous for its 'lyre-shaped' tail and vocal mimicry. **IC**

The Father of Nepalese Botany

1802–1803

Although Dr Francis Buchanan-Hamilton (1762–1829) was one of the most prolific natural historians ever to work in the Indian subcontinent, his discoveries mainly became known through the work of later authors. During two decades of employment by the British East India Company, this Scottish surgeon and statistical surveyor explored an exceptional geographical range in what are now the countries of Myanmar, Bangladesh, Nepal and India. His earliest collections and plant descriptions, given to the famous botanist Sir Joseph Banks, had been left unused or misattributed. In 1806, Buchanan-Hamilton presented his extensive descriptions and drawings from South India and Nepal to a friend from his Edinburgh University days, Linnean Society founder James Edward Smith.

During 1800–1801 (a period of political upheaval in South India following the fourth Anglo-Mysore War), Buchanan-Hamilton described over 300 plants from the region, and over 1,200 from Nepal in 1802–1803, two-thirds of which he believed to be new to science. These were the first scientific collections from this secretive Himalayan country, and why he is recognised as the 'father of Nepalese botany'. The Mysore and Nepalese specimens were accompanied by meticulous Latin descriptions and exquisite coloured drawings, most likely the work of Haludar, an Indian artist who accompanied him from Bengal. As seen here, the specimens, manuscripts and drawings are still in the Society's collections, including 100 plant drawings each from Mysore and Nepal, together with a small number of earlier paintings of plants and fish from East Bengal.

Smith, like Banks, did little with Buchanan-Hamilton's collections, publishing a mere handful of new species and, worse, preventing access to the material. However, after 200 years of obscurity, research on the Society's Buchanan-Hamilton collections is at last revealing their scientific and cultural importance. **MW**

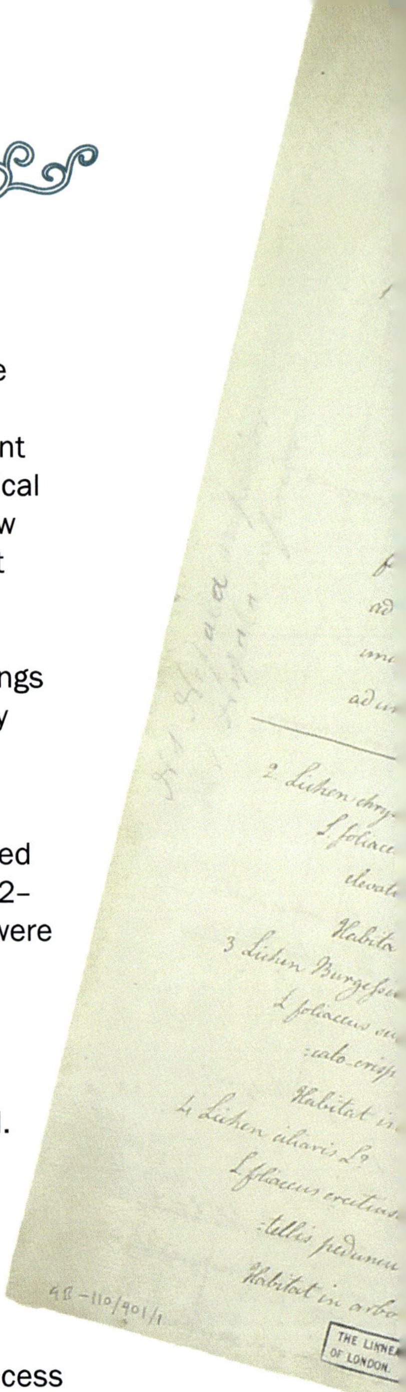

Flora Nepalensis
disposita
Ordo I. Fungi
237
Tab.
16
Upper Nepal
Nepal.

4
5
7
Fig: 1
9

William Roxburgh's Silk Worms

1802

Botanical study was crucial for a surgeon in the 18th century when William Roxburgh (1751–1815) was at Edinburgh University. Roxburgh, known as the 'founding father of Indian botany', arrived in South India as a doctor in the British East India Company in 1776, and set about gathering material to send home (his collecting actually extended across India, the East and the Americas). On occasion, he was accompanied by Johann Gerhard König (1728–1785), once a pupil of Linnaeus, and the naturalist for the Company. Plant specimens were sent to both Sir Joseph Banks at the Royal Botanic Gardens, Kew and James Edward Smith at the Linnean Society. As few live plants survived the months-long journey round the Cape of Good Hope, Roxburgh organised Indian artists to draw and paint the plants, as well as sending seeds home. Sadly these artists remain unknown, though their work introduced many Indian species to British naturalists. Over the years, more than 2,500 drawings were produced and 1,000-plus species sent to Kew; out of this came *Plants of the Coast of Coromandel* (1795–1820) and *Flora Indica* (1820–1824).

Profit was a driving force for the British East India Company and its employees, which meant that Roxburgh sought plants with economic potential, e.g. pepper, sugar, hemp and dyes. Roxburgh worked on teak; wood and hemp were necessary for the rigging of naval ships during the Napoleonic War. (He was also doing surveys into the possibility of building canals for irrigation.) In 1793 he became the first paid Superintendent of the Calcutta Botanic Garden, which became of world-class importance. Roxburgh continued to look for commercially important species; he tried growing tea but realised that the Calcutta climate was not suitable, and attempted to grow the cactus *Opuntia* (the host plant for the cochineal beetle).

He studied the silk worm and the possibility of developing a local silk trade, eventually sending a description and drawings to Smith at the Linnean Society. Entitled 'Account of the Tusseh and Arrindy Silk Worms of Bengal', it was read at the Society in 1802 and published in its *Transactions* in 1804. **TR**

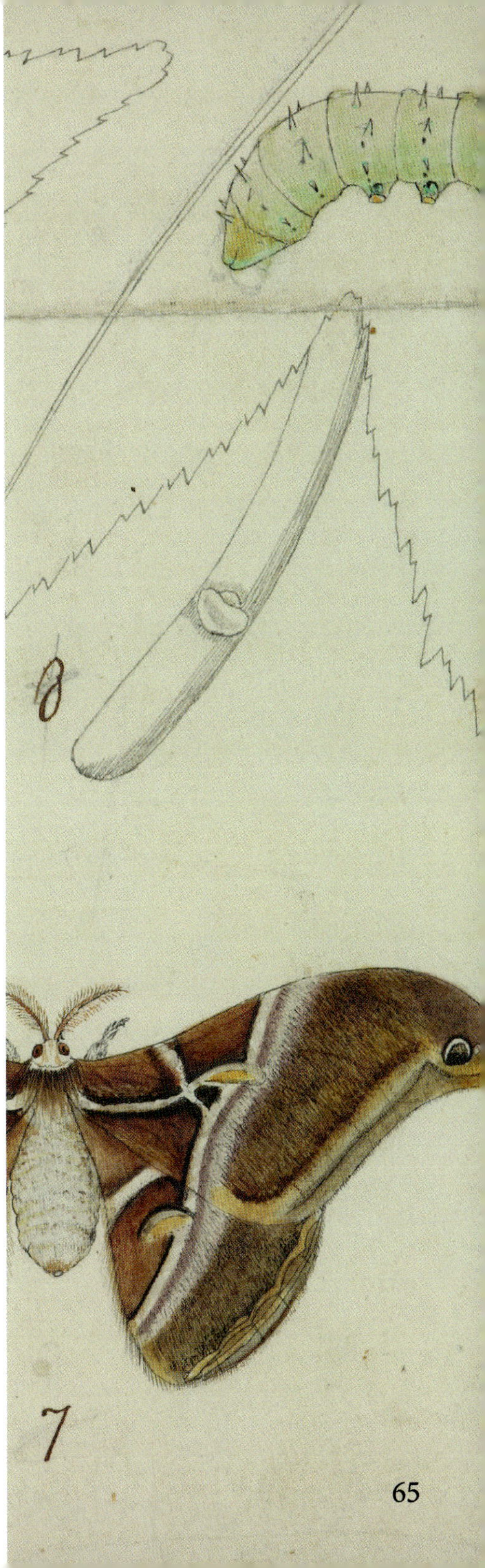

Flora Graeca: Accuracy & Elegance

1806–1840

The late 18th and early 19th centuries saw a burgeoning of published works and letters on natural history, including mineralogy and meteorites. The formation of societies like the Linnean, the gentlemanly grand tours, exploration of distant lands, and the active collection of specimens, all provided continuing stimuli for research and communication to a growing audience.

James Sowerby (1757–1822—Top Right), with his remarkable creative energy and skill, was one of the most important artists and communicators of his day. He was the patriarch of a family that produced several generations of natural history illustrators. He worked with botanist William Curtis (1746–1799), contributing fine engraved plates for the magnificent *Flora Londinensis* and the popular *Botanical Magazine*, and with the Society's founder James Edward Smith on the *Coloured Figures of Rare Plants* (containing 'some of Sowerby's most accomplished work') and the famous *English Botany* of some 36 volumes. His illustrated works covered fungi, crystals, minerals, meteorites, fossils and some British animals, as well as a quantitative study of light and colour. He named numerous new species, including his 'own' Sowerby's Whale. His trademarks were accuracy and elegance; authors sometimes wrote that there was no need for a written taxonomic description as Sowerby's engraving was sufficient in itself.

His name is also strongly linked to one of the most famous (and possibly expensive) botanical works produced at this time—John Sibthorp's *Flora Graeca* (1806–1840), the product of Sibthorp's travels and research in Greece, published almost immediately before his death. Smith was appointed as 'author' of the science while Sowerby was enlisted to engrave the plates from fellow artist Ferdinand Bauer's original drawings (*Dracunculus vulgaris*—Opposite). This rare work is considered as one of the finest examples of botanical illustration and the Linnean Society is proud to hold it in its collections. **PH**

30.
Achatina marginata
30
Pl 49
Oxyrhynchus.
49
Plate 125.
No 28. November 1822.
Papilio Nireus.
125
Zool. Ill. Vol 3.
Pl 141
No 29. February 1823
86

Colouring Book

1820–1823

William Swainson (1789–1855) was a naturalist interested in the study of zoology, namely birds, molluscs and insects. He was also an accomplished artist, as the treasure shown here testifies. Born in London, Swainson was the son of one of the Society's first Fellows, John Timothy Swainson (1756–1824), who held prominent positions in customs in both London and Liverpool. A speech impediment made William's education difficult, and at the age of 14 he became a clerk at the Liverpool Customs. A turn in the Army Commissariat led him to Sicily which sparked an interest in ichthyology, and a trip to Brazil saw him return to England laden with thousands of insect and plant specimens, bird skins and drawings.

His interest in zoology impelled him to publish several works, including the highly regarded *Fauna Boreali-Americana* in 1831. Our interest here relates to possibly his best-known work, *Zoological Illustrations* (1820–1823). The Society holds a wonderful set of proof or 'pattern' plates for the book, which spans entomology, conchology and ornithology. Swainson was the pioneer of lithography for natural history illustration in Britain; *Zoological Illustrations* is the forerunner of the great illustrated folios of Edward Lear and John and Elizabeth Gould.

In lithography, the image is drawn on a block of limestone with a greasy crayon. The stone is then chemically treated so that in printing, ink adheres to the lines where the original drawing had been made and is repelled by the dampened surface of the porous stone. Though cheaper than copperplate printing, the resulting impressions still needed to be coloured by hand, using pattern plates as a guide (in this case, around 225,000 individual plates needed to be coloured). The pattern plates for the colourists to follow were coloured and annotated by Swainson himself. **LB & RG**

Thomas Hardwicke's Zoological Drawings

1821

A career soldier in the British East India Company, Thomas Hardwicke (1756–1835) arrived in Calcutta in 1778, rising to Major-General in 1819. A passionate naturalist, in his spare time he recorded the beautiful and bizarre species he found on his travels, regularly submitting papers to scientific societies in Calcutta and London.

This image of a reticulated python (*Python reticulata*), dated December 1821, was created to accompany one such paper, sent to the Linnean Society in 1823. The animal was drawn from life by an unknown Indian artist, trained in the schools that had sprung up in Calcutta to paint for European patrons. Hardwicke kept his own private menagerie, and the python appears to have been a prized exhibit. In his meticulous way, he noted that it ate '4 chickens, two Pigeons, two Rats, and one crow'.

His painting of the red panda (*Ailurus fulgens*) was drawn in 1820 from pelts received from Nepal via Nathaniel Wallich. Knowing it by the Nepali name, 'Wah', Hardwicke mistakenly believed the animal to be a marten (*Mustella zebellina*) and sent an article about it to the Linnean Society in 1821. Unfortunately, by the time the paper was published in 1827, a French zoologist, Frédéric Cuvier, had named the species himself, based on fragmentary material given by Hardwicke to Alfred Duvaucel. Hardwicke's painting, likely by J. Hayes, an otherwise unknown British or Anglo-Indian natural history painter, is the first image of a red panda ever to have arrived in Europe.

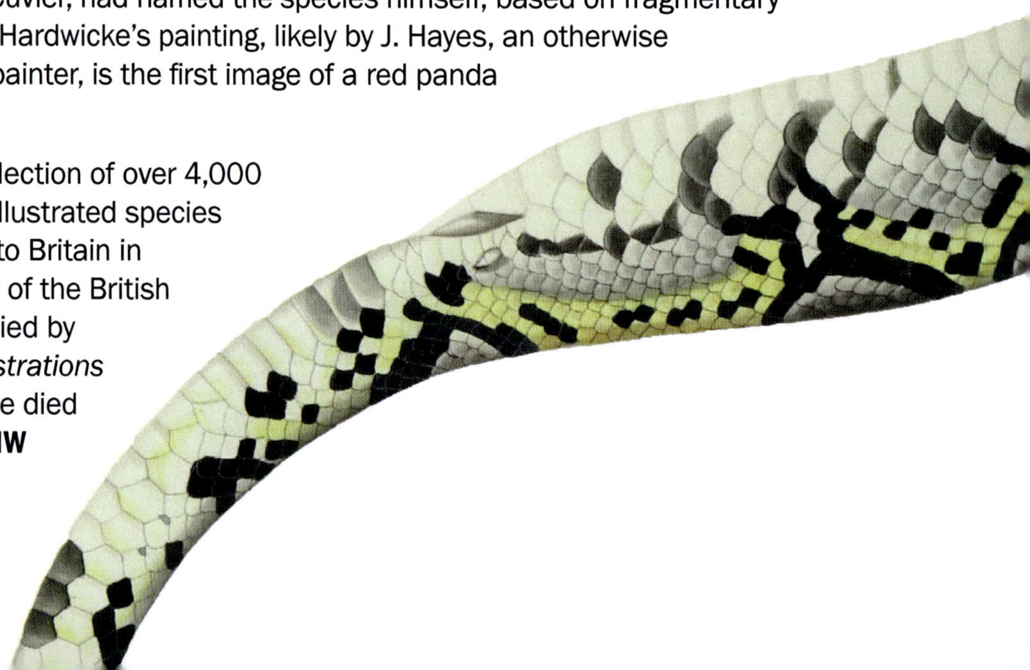

Hardwicke eventually amassed an important collection of over 4,000 paintings of animals and plants, many of which illustrated species unknown to European naturalists. On his return to Britain in 1823, Hardwicke approached John Edward Gray of the British Museum with a plan to publish them, accompanied by Gray's scientific descriptions. The result was *Illustrations of Indian Zoology* (1830–1835). Sadly, Hardwicke died before further volumes could be finished. **DL & MW**

Robert Brown's Microscope

1827–1833

A brown-paper parcel was delivered to the Society in January 1922. Inside was a small brass microscope in a mahogany box, and a letter which read: 'I have much pleasure in offering Mr Brown's microscope to the Linnean Society if they care to accept it ... its history since its original owner is accounted for. Yours faithfully, Ida M. Silver (Miss).'

This microscope had been owned by Scottish botanist Robert Brown (1773–1858), famous for observing the ceaseless jostling of microscopic particles in 1827 (due to molecular movement) that we now call Brownian motion. His later research, using orchid tissue, also gave us the concept of the cell nucleus, and the resulting paper 'On the Organs and Mode of Fecundation in Orchideae and Asclepiadeae' was published in the Society's *Transactions* in 1833.

Surprisingly, the Fellows of the Linnean Society had dismissed it as 'a simple dissecting microscope', convinced that it could never have been used for such serious scientific research. Some botanists, including Irene Manton FRS (1904–1988), realised that the lens of this little microscope could reveal more than previously thought, and it was decided in 1981 that Brian J. Ford should restore it to working condition.

Once it had been carefully cleaned and re-assembled, its tiny hand-ground lenses could bring the microscopic world to life. Yes, cells could be clearly observed, and even the cell nucleus—which had been believed too small for Brown to observe—could be studied. Made by Robert Bancks around 1810, the microscope relied on a single magnifying lens. Charles Darwin was impressed by what Brown showed him and was recommending this kind of microscope for field use many decades later.

The gift that Miss Silver offered to the Society may have been disregarded as unimportant a century ago—but today we can recognise just how powerful, and important, this beautiful little microscope really is. **BJF**

William Carey's Indian Insects

1828

On 4 November 1828, the Linnean Society received a packet of 35 watercolour drawings from Bengal. It was a gift from one of its most distinguished Fellows, the Rev. Dr William Carey (1761–1834), one of the greatest of all Orientalists. From a humble background, he started his career as a shoemaker before becoming a Baptist missionary; in 1793 he went to Bengal, where he spent the rest of his working life.

As a boy in his native Northamptonshire, England, Carey had been interested in insects and their metamorphoses, which explains the inclusion of 22 watercolour plates showing the life cycles of various insects with their host plants. Some show crop plants such as jute, mustard and taro with their associated pests, but in some cases the unknown artist may have confused the host plant and related insect species. Additionally, the limited annotations on the drawings make identification quite difficult, especially the 13 plates that show numerous species of insects of groups including beetles, cicadas and grasshoppers.

Carey's linguistic skills were remarkable, throwing himself into the study of Bengali and Sanskrit with the help of Brahmin pandits (scholarly priests) and the great Sanskrit scholar Henry Thomas Colebrooke, who in 1823 proposed him for Fellowship of the Linnean Society. Carey became Professor of Bengali and Sanskrit at the College of Fort William in Calcutta, but his main base was the nearby Danish colony of Serampore, where in 1800 he had joined the missionaries William Ward and Joshua Marshman. Together they set up a printing press to print Bible translations and dictionaries for numerous Indian languages, but also the pioneering botanical works of William Roxburgh—*Hortus Bengalensis* in 1814 and *Flora Indica* in two editions between 1820 and 1832. **HN**

To knit a childs
Cast on 36 - Waist -
This from the back
Then cast on 68
Then 36 for the other
Then 38 times 4
Myrtle Leaf
Shell Pattern
57 Stitches -
Knit a plain
Cast on one row - 46
Knit 8 loops into 1
Then knit 16 loops into 8
23 Stitches make one pattern
eleven
Sleeves
times 4 -
Lady Smith
Lowestoft

An Influential Victorian Woman

1828–1878

Lady Pleasance Smith (1773–1878) was the wife of the Linnean Society's founder, James Edward Smith. After her husband's death in 1828, Pleasance maintained a lively correspondence with more than 100 different people, resulting in over 550 surviving letters addressed to her that reside in the Society's archives today. Amongst her correspondents were doctors, writers, scientists and noted theologians. Yet her link to the Society did not fade with Smith's death; she communicated with members throughout her long life.

Pleasance was known for her philanthropy and intelligence, which prompted several of her correspondents to pen poetry about Pleasance and Smith. In addition to the letters, ephemera (Left) including knitting patterns, riddles and recipes reveal glimpses of a bright, generous woman whose wide network of acquaintances sought her opinion. The Society's collections also contain two beautiful portraits of a youthful Pleasance by John Opie, one of which is shown here.

Although there are no letters in her own hand, the correspondence sent to Pleasance demonstrates her interest in wide-ranging subjects, from arts and humanities to sciences and the natural world. On the whole, these letters reflect the concerns of the time: war, natural disasters, British East India Company news and personal connections, including encounters with contemporaries like Charles Darwin, Charles Dickens and Queen Victoria. They spotlight the fascinating life of an influential Victorian Lady, who actively shaped the world around her. **TH & KW**

Edward Lear's *Parrots*

1830

English painter and poet Edward Lear (1812–1888) is best known for his nonsense literature, most famously 'The Owl and the Pussycat', but he began his career as a brilliant natural history illustrator. Specialising in ornithological illustration, Lear was first contracted by the Zoological Society of London but was later hired in the 1830s by Edward Smith Stanley, 13th Earl of Derby (1775–1851) and President of the Linnean Society (1828–1834), to paint the rarer species in his large menagerie. Produced for private distribution as *Gleanings from the Menagerie and Aviary at Knowsley Hall* (1846), a copy of this rare folio sits in the Linnean Society's library.

Lear drew from live subjects, many of which he found in the Zoological Gardens in London's Regent's Park, and it could be argued that this shows in his illustrations. One of his most renowned publications, and our object of note, was his beautiful monograph, *Illustrations of the Family of Psittacidae, or Parrots* (1830–1832), which he began publishing in parts in 1830; incredibly, he was only in his late teens at the time. (Lear was elected as an Associate Member of the Linnean Society that same year.) Considered a superb blend of artistic merit and scientific study, it was well received and placed the young Lear in the same category as respected illustrators like John James Audubon and John Gould. Referring to the famous scarlet macaw (*Ara macao*—LEFT), ornithologist William Swainson wrote that it was equal to anything by 'Audubon, for grace of design, perspective, or anatomical accuracy'.

The colours in his illustrations are sharp and bright; Lear used strong opaque watercolours, and sometimes included a veil of egg white to bring life to the feathers. The Society holds a beautifully preserved copy of the first edition, shown here. **LB**

Charles Darwin's Voyages of Discovery

1831 & 1881

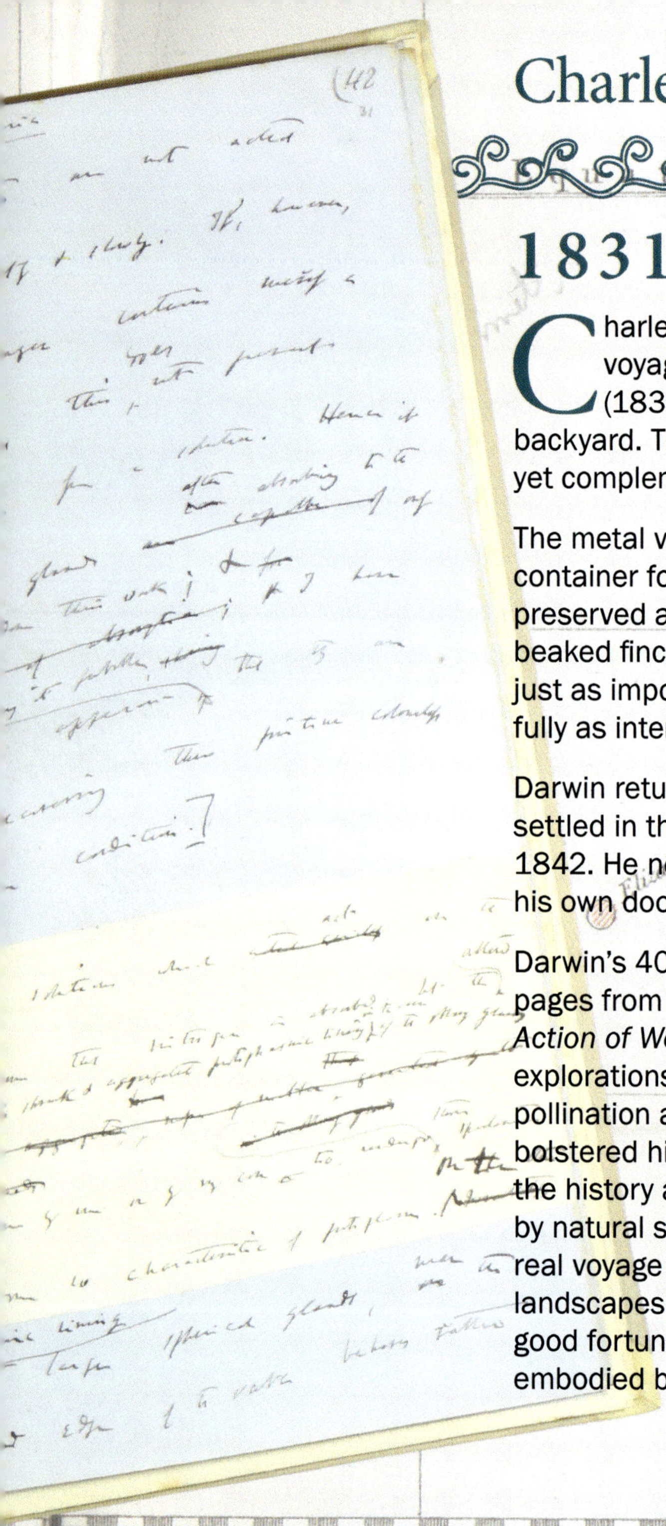

Charles Darwin (1809–1882) embarked upon not one, but two great voyages of exploration: his global, five-year expedition on HMS *Beagle* (1831–1836) and the near 40-year journey of discovery in his own backyard. The artefacts highlighted here are windows into these very different, yet complementary voyages.

The metal vasculum (Opposite) was used by Darwin on the *Beagle*; a protective container for collecting plant specimens in the field until they can be properly preserved and pressed. While the charismatic giant tortoises and variously-beaked finches of the Galápagos Islands loom large in Darwin lore, plants were just as important in the development of his thinking: 'The botany of this group is fully as interesting as the zoology', he wrote in his *Voyage of the Beagle*.

Darwin returned home in October of 1836, later married Emma Wedgwood and settled in the Kentish village of Downe with his young and expanding family in 1842. He never travelled to distant lands again, but instead explored nature on his own doorstep, in fields and in meadows.

Darwin's 40-year odyssey with earthworms is reflected in these manuscript pages from his 1881 book, *The Formation of Vegetable Mould, Through the Action of Worms* (Left). His studies bookended a seemingly endless array of explorations, ferreting out the secrets of barnacles and bees, pollination and pigeons, weeds and worms. All of these data bolstered his astonishing (and, to some, incendiary) idea that the history and diversity of life on Earth is a tale of evolution by natural selection. Marcel Proust once wrote that 'the real voyage of discovery consists not in seeking new landscapes but in having new eyes'. Darwin did both, to our good fortune; literal and figurative voyages of discovery embodied by these fascinating artefacts. **JTC**

Mora Tree. By Mr. Robert H. Schomburgk.
cated by George Bentham, Esq. F.L.S.
Read March 20th, 1838.
NT among the trees which adorn the f
nish by their profuse verdure and giganti
e king of the forest. Rising to the heigh
t gives out its branches, it towers ove
the banks of the rivers of Guiana, for
ge, overshadowing numerous minor t
the form of natural festoons. The t
nts towards the base into tabular b
property the silk-cotton tree, Bo
an early decay, and they some
protection against the incl
often wondered, when
excrescences had to
so frequently sw
The Mora, o
for naval architect
introduced for building
apprehension that there
plied. The wood is une
when driven into it: w
Mora excelsa.

The Stunning Work of Miss Drake

1838

During a prolific, though relatively short career of less than 20 years, Sarah Ann Drake (1803–1857) had the distinction of working as a botanical illustrator with many of the great names of the age, including Nathaniel Wallich, John Forbes Royle, William Jackson Hooker and James Bateman. Most prominently, she produced more than 1,100 illustrations under the employ of the English botanist and celebrated orchidologist John Lindley (1799–1865). Written documentation relating to Miss Drake, the cognomen with which she chose to sign her paintings, is scarce, and mostly relates to her working and familial relationship with her chief employer, Lindley, who named the genus *Drakea*, the hammer orchids, in her honour.

Illustrations like this drawing of *Mora excelsa* exhibit evidence of an informed comprehension of botany; in the accompanying text in the Linnean Society's *Transactions* of 1838, explorer Robert H. Schomburgk (1804–1865) repeatedly refers to the *Mora* as 'gigantic', but it was essential for the figure to be represented accurately and to scale, despite the restrictive size of the journal's octavo format.

Drake developed proficiency in working from dried specimens to create expressive and accurate illustrations, as access to recently-discovered live plants would have been sporadic. The drawing displays astonishing clarity conveyed through Drake's modelling in pencil and wash, and although she originated many colour plates, it is this format in which her genius is most fully articulated. **LP**

Anna Atkins: Picture Pioneer

1843

The 19th century witnessed leaps forward in both printing techniques and the art of photography. One such leap occurred in the work of English botanist Anna Atkins (1799–1871), regarded by many as having produced the first photographic book. Her work even predates *The Pencil of Nature* (1844) by revolutionary photographer William Henry Fox Talbot by some eight months.

Atkins had grown up around science (her father, John George Children, was a chemist, mineralogist and amateur zoologist), and she was inspired by polymath Sir John Herschel's work with cyanotypes, whereby a specimen is placed on paper treated with potassium ferricyanide and ferric ammonium citrate, and placed in direct sunlight. The paper reacts, turning blue, but stays white where the specimen is positioned, creating a print of the species. Dedicated to her father, Atkins published her pioneering work, *Photographs of the British Algae: Cyanotype Impressions* (1843), using the cyanotype process. And because of the very individual nature of the printing, each copy is slightly different.

Why algae? Atkins was active during a time when there was a surge in the study of natural history in Britain, and algae was attractive and relatively easy to source. Atkins had been studying other algal works but was unimpressed with the amount, or detail, of the illustrations, and in response she produced *Photographs of the British Algae*. Produced in sections, it was up to her lucky acquaintances (the recipients of her work) to have them properly bound. Later, she published *Cyanotypes of British and Foreign Ferns* (1853) and *Cyanotypes of British and Foreign Flowering Plants and Ferns* (1854).

Subsequently largely forgotten, her work was revived and celebrated in the 1970s by American art historian Larry Schaaf. It is believed that Atkins produced only 17 copies of *Photographs of the British Algae*, and we have been told by enthusiasts that the Society's copy is beautifully preserved. **LB**

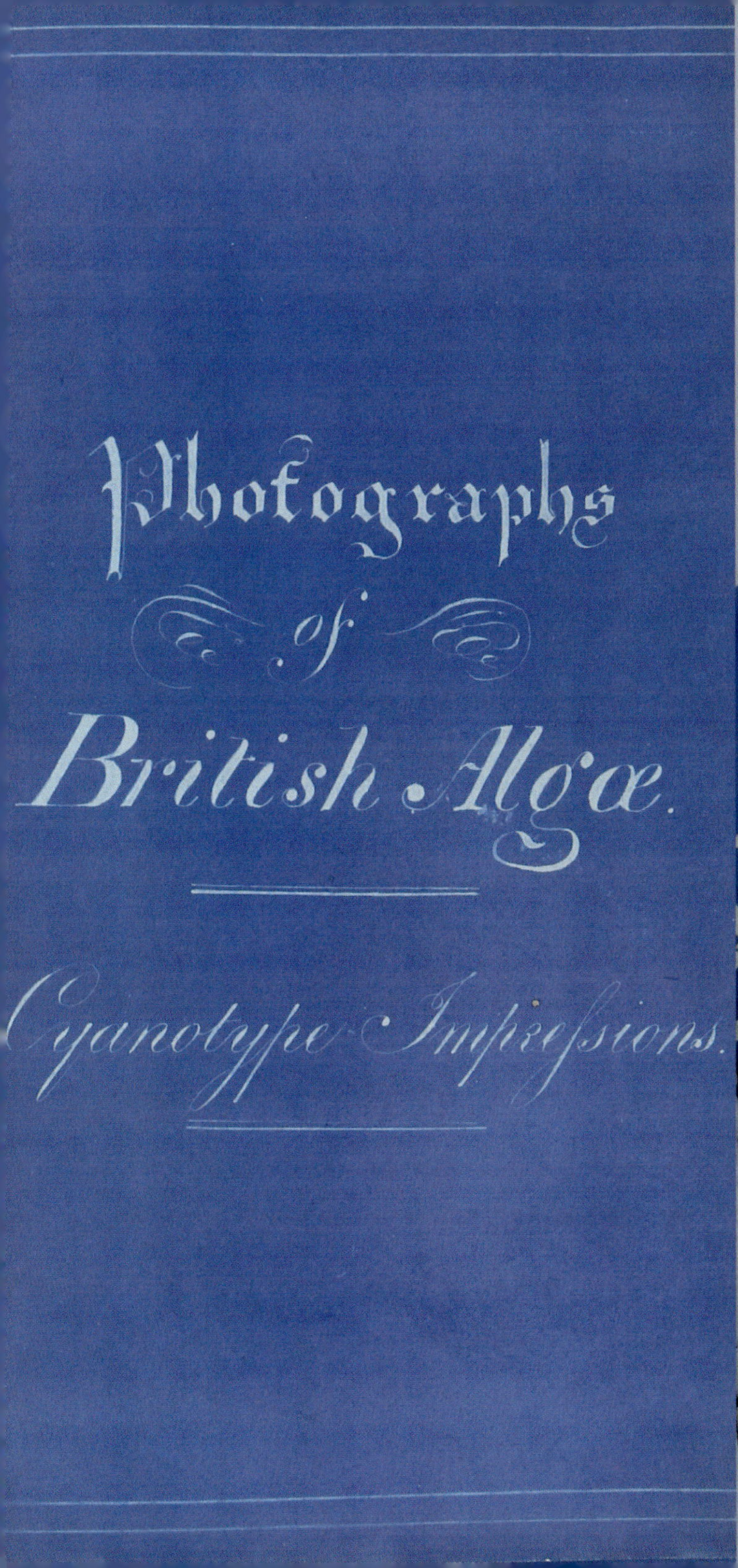

Photographs of British Algæ.

Cyanotype Impressions.

Alfred Russel Wallace's Palms of the Amazon

1852

Naturalist Alfred Russel Wallace (1823–1913) scrambled into his smoke-filled cabin and hastily stuffed a few items into a small tin box, including the pencil illustrations of palms and fishes he had made during his four-year expedition to the Amazon. It was 6 August 1852, and his ship was on fire in the mid-Atlantic. All his natural history collections in the hold would soon be destroyed, but he and the crew would miraculously survive, rescued 10 days later in their lifeboats, by a cargo ship on its way to England.

Back in London, Wallace had to make the most of the meagre materials he had salvaged. He wrote two books: one about his travels, and the other about Amazonian palms and their ethnobotanical uses. The latter was the first ever field guide to palms, describing 12 new species including *Leopoldinia piassaba* (BELOW), the source of a fibre used for brooms, ropes and other items in both Brazil and Britain. The manuscript of the book is now cared for by the Linnean Society. **GWB**

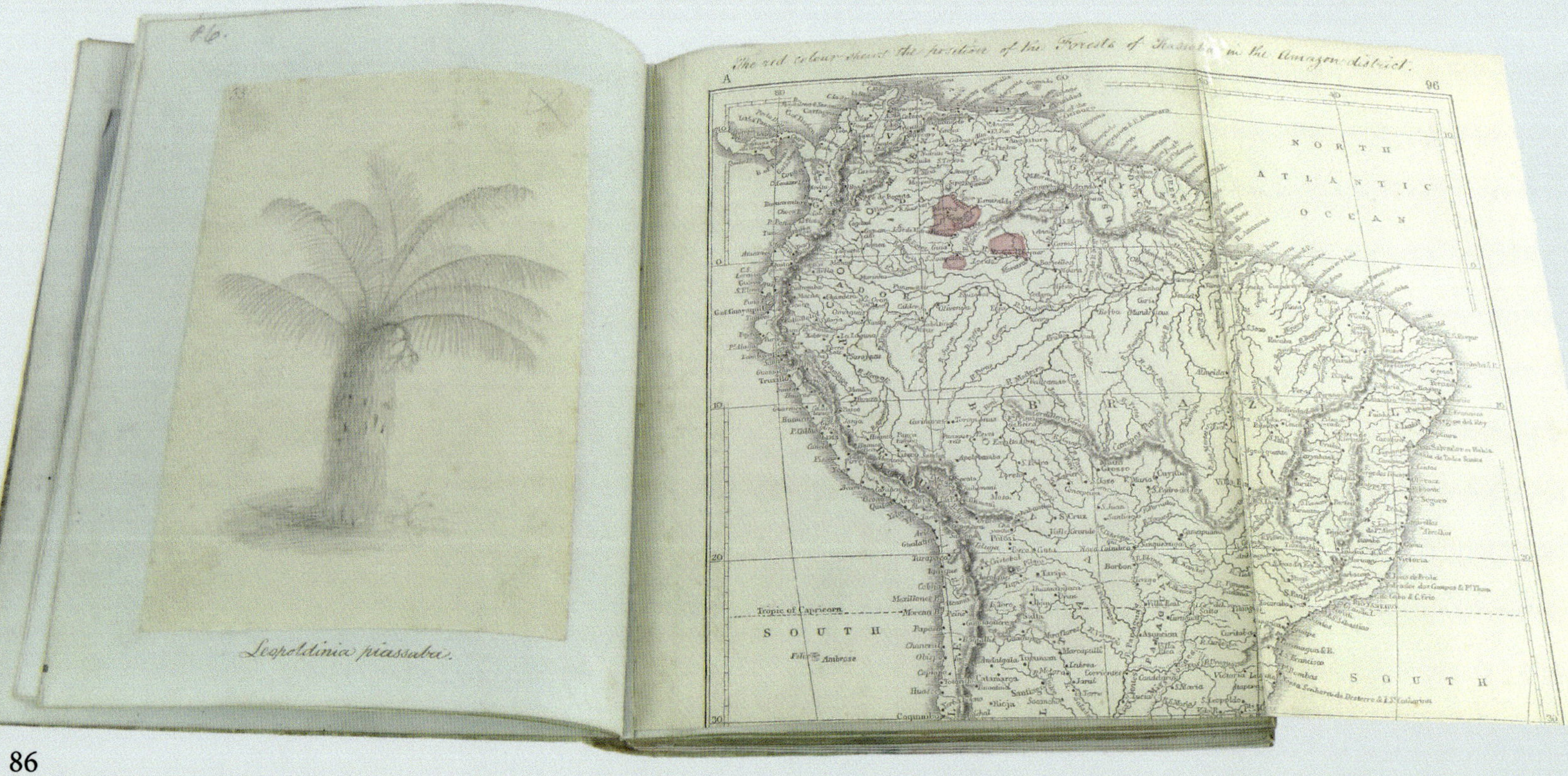

The Malay Archipelago & Python Skin

1854–1862

Wallace described his eight years in the Malay Archipelago (between 1854–1862) as 'the central and controlling incident of my life'. He travelled extensively around what is now Singapore, Malaysia, Indonesia and East Timor, collecting animal specimens, observing the indigenous peoples and deciphering the workings of evolution. His collections included 126,000 specimens, including some 5,000 new species. Not only did he document the invisible barrier separating the Oriental and Australian zoogeographic regions (dubbed the Wallace Line), he also discovered natural selection, publishing the idea with Charles Darwin in 1858.

Wallace's famous travelogue, *The Malay Archipelago* (1869), was his most successful book; it has never been out of print and has been translated into many different languages. One of its illustrations shows a 12 foot long python being unceremoniously evicted from Wallace's hut in Ambon; the python skin is housed at the Linnean Society. **GWB**

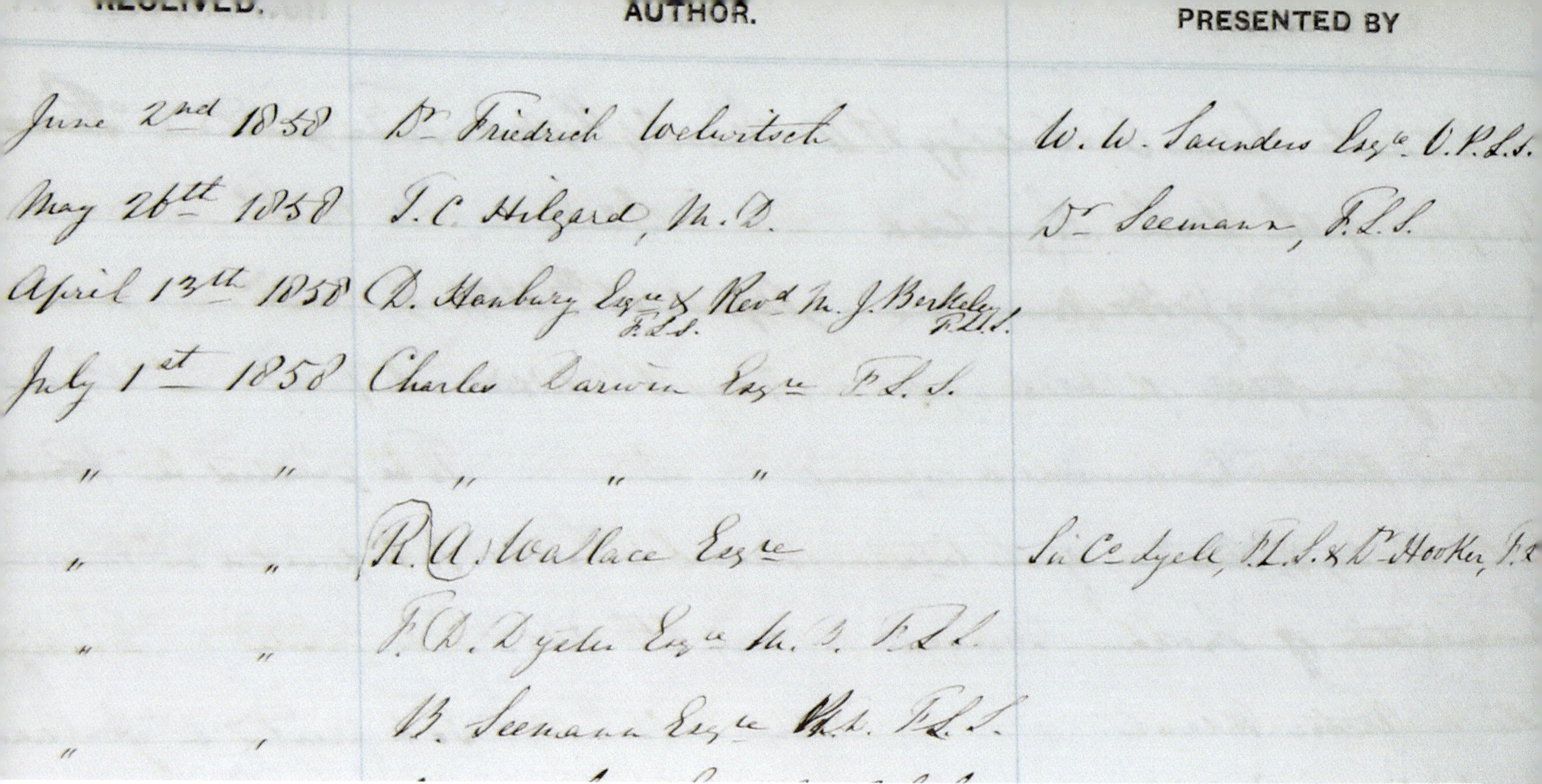

RECEIVED.	AUTHOR.	PRESENTED BY
June 2nd 1858	Dr Friedrich Welwitsch	W. W. Saunders Esq. V.P.L.S.
May 26th 1858	J. C. Hilgard, M.D.	Dr Seemann, F.L.S.
April 13th 1858	D. Hanbury Esq. F.L.S. & Revd M. J. Berkeley F.L.S.	
July 1st 1858	Charles Darwin Esq. F.L.S.	
" "	" " "	
" "	A. R. Wallace Esq.	Sir C. Lyell, F.L.S. & Dr Hooker, F.L.S.
" "	J. D. Dyster Esq. M.D. F.L.S.	
" "	B. Seemann Esq. Ph.D. F.L.S.	

The Darwin-Wallace Papers on Evolution

1858

On 1 July 1858, an event took place that would change biological science and the way we view the natural world forever. That day, at a meeting held at the Linnean Society, Charles Darwin and Alfred Russel Wallace's seminal papers proposing evolution by natural selection were read before a small audience. This breakthrough in our understanding of life on Earth met with little reaction at the time, but the 'joint paper' did not go unnoticed after it was published in the Society's *Journal* in August of that year (Opposite Right). Historian Janet Browne notes: 'During the next two or three months it was reprinted either in full or in part in several popular natural history magazines... A number of people made their views known in letters, reviews, and journals.' However, when Darwin's book On the *Origin of Species* appeared in November 1859, it popularised the theory and became the focus of debate. For Wallace

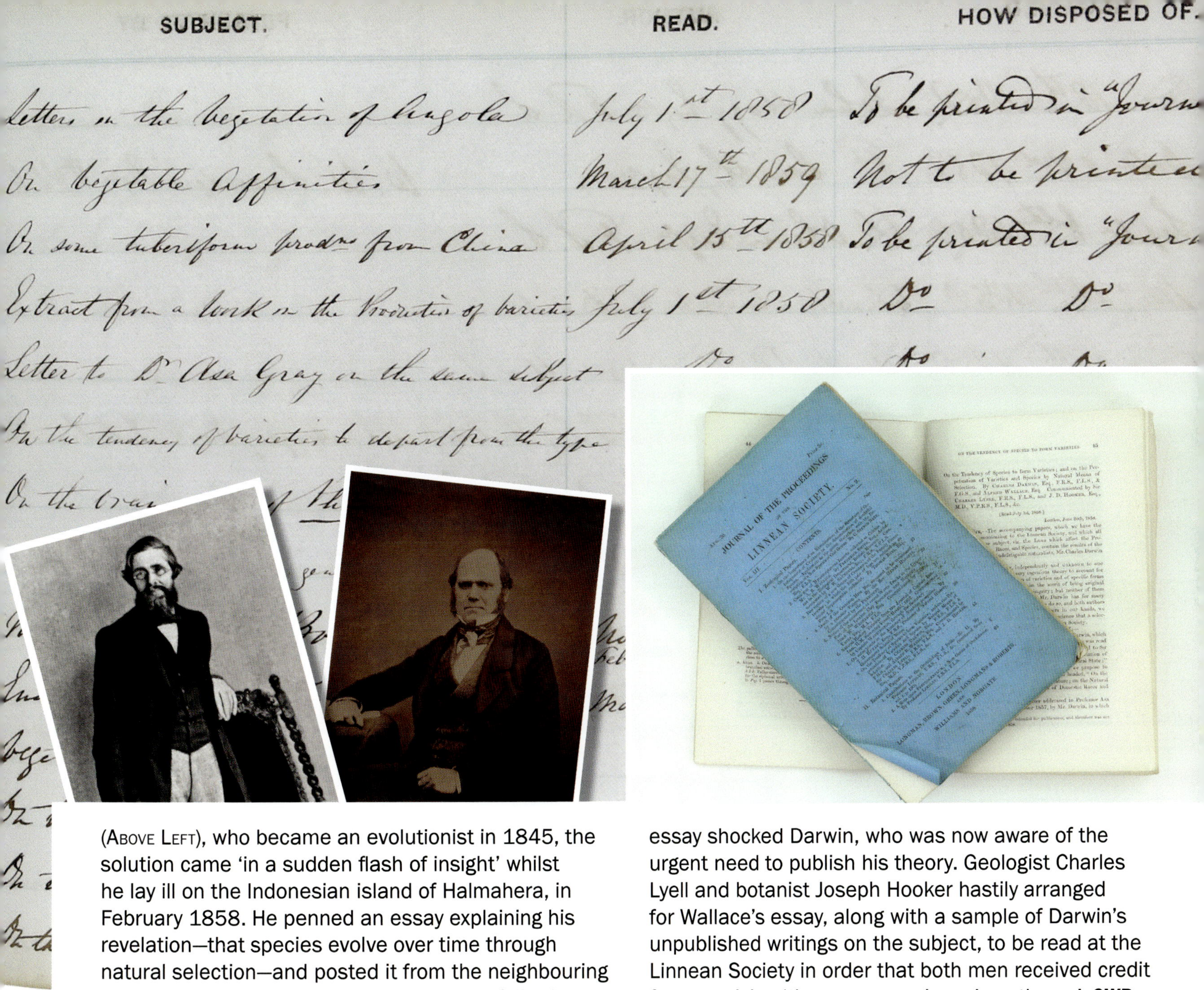

SUBJECT.	READ.	HOW DISPOSED OF.
Letters on the Vegetation of Angola	July 1st 1858	To be printed in "Journ
On Vegetable Affinities	March 17th 1859	Not to be printe
On some tuberiform prodns from China	April 15th 1858	To be printed in "Journ
Extract from a work on the Production of varieties	July 1st 1858	Do Do
Letter to Dr Asa Gray on the same subject	Do	Do Do
On the tendency of varieties to depart from the type		
On the		

(Above Left), who became an evolutionist in 1845, the solution came 'in a sudden flash of insight' whilst he lay ill on the Indonesian island of Halmahera, in February 1858. He penned an essay explaining his revelation—that species evolve over time through natural selection—and posted it from the neighbouring island of Ternate to Darwin in England the following month. Little did he know that, 20 years earlier, Darwin (Above Right) had arrived at the same idea. Wallace's essay shocked Darwin, who was now aware of the urgent need to publish his theory. Geologist Charles Lyell and botanist Joseph Hooker hastily arranged for Wallace's essay, along with a sample of Darwin's unpublished writings on the subject, to be read at the Linnean Society in order that both men received credit for conceiving 'the same very ingenious theory'. **GWB**

Dr Shortt's Birds of Southern India

1859

Open a palm-sized album in the Linnean Society library and be dazzled by the jewel-like colours of two dozen bird paintings, glowing from sheets of mica, as bright as the day they were painted in South India in the 1850s. The album was presented to the Society in 1859 by Dr John Shortt (1822–1889), a British East India Company surgeon, probably as a 'calling card' shortly before his election as a Fellow. Shortt was Anglo-Indian, having trained as an Apothecary at the Madras Medical School. As a reward for saving the life of a wealthy cholera-stricken army officer, he was enabled to travel to Britain where he obtained an MD from King's College, Aberdeen. This allowed Shortt to enter the Company's Medical Service, where he also pursued wide-ranging natural history and anthropological studies, some of which were published in the Society's journals.

Twenty of the paintings in the album are versions of ones commissioned by another Company surgeon and FLS, Thomas Caverhill Jerdon (1811–1872), and published in his *Illustrations of Indian Ornithology*. At this time, Trichinopoly (now Tiruchirappalli) was a well-known artistic centre in what was then the Madras Presidency where Jerdon trained artists to make accurate bird paintings. The paintings may have been copied from Jerdon's book; however, as four are not in the book, and some of the identifications are incorrect, the artist may have kept a stock of images from which to make versions for later patrons. Although Shortt believed all the birds to be South Indian, some of the specimens depicted were sent to him from the Himalayas by the renowned zoologist Edward Blyth of Calcutta (1810–1873). **HN**

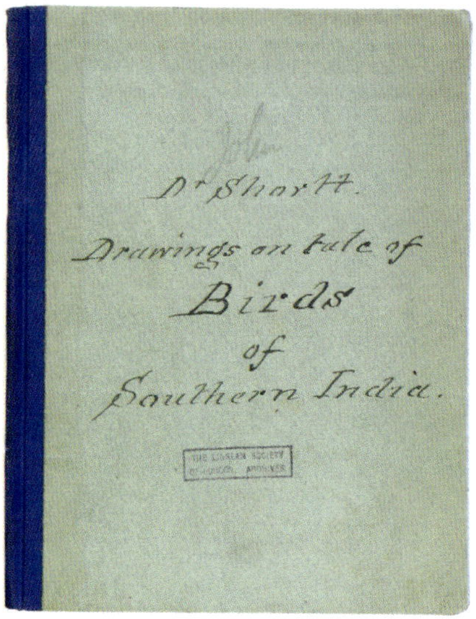

Darwin's 'Best' Portrait

1881

Hanging in the Meeting Room of the Linnean Society is an intimate and moving portrait of Charles Robert Darwin as an old man, painted a year before his death. Artist John Collier (1850–1934) depicts Darwin wearing a loose cloak and holding a hat in his left hand. He faces the observer whilst emerging from atmospheric obscurity, his expression sombre but complex.

Within the Society's correspondence is a letter from Darwin, dated 27 May 1881, which tells the story of his portrait, as commissioned by the Society. Writing to the Society's Zoological Secretary George John Romanes (1848–1894), Darwin expresses his reluctance to sit for the portrait: 'It tires me a great deal to sit to anyone, but I sh[oul]d be the most ungrateful & ungracious dog not to agree.'

However, he agreed and recommended his choice of artist, based on his admiration of Collier's portrait of Darwin's friend, Joseph Dalton Hooker: 'If I am to sit, it would be a pity not to sit to a good artist, & from all that I have heard I believe Mr Collier is a very good one.'

According to Darwin's third son, Francis: 'Many of those who knew his face most intimately, think that Mr Collier's picture is the best of the portraits', a fact that even Darwin himself agreed with. Copies of the portrait hang at Darwin's family home of Down House in Kent, and at London's National Portrait Gallery, but the Society's is the original. **VM**

May 27th 1881

DOWN,
BECKENHAM, KENT.
RAILWAY STATION
ORPINGTON. S.E.R.

My dear Romanes

I thank you for your kind & very considerate letter. I have not the least idea, who is the artist of whom Mr Murie has thought. It tires me a good deal to sit to anyone, but I shd be the most ungrateful & ungracious dog not to agree cordially, supposing that enough is subscribed, about which

INDIES.
NEW
ZEALAND.

Clever Useful Pets

1896

Zoologist Albert K. L. G. Günther's (1830–1914) series of 42 albums show the world's fauna, and offer a much more visual and personal experience than his impressive legacy of published works. The heavy red bindings struggle to contain an explosion of colour from the original drawings, historical prints, proof copies and early photographs contained inside.

The term 'albums' seems to indicate they were created as a whimsical pastime, but the many cross-references and accompanying notes suggest nothing could be further from the truth. As the donations register of the Linnean Society clearly states, they were Günther's workbooks.

Günther, Keeper of Zoology at London's Natural History Museum (1875–1895) and President of the Linnean Society (1896–1900), ordered the volumes geographically, then divided them into animal classes. They boast the work of British artists like George Edwards, as well as artists from around Europe, and Japanese Hokusai woodblock prints (Opposite). One particularly poignant aspect is the presence of rare and extinct species, often in original illustrations—the dodo, the giant moa of New Zealand, and Northern Europe's very own original 'penguin', the great auk (*Pinguinus impennis*—Right), to name a few.

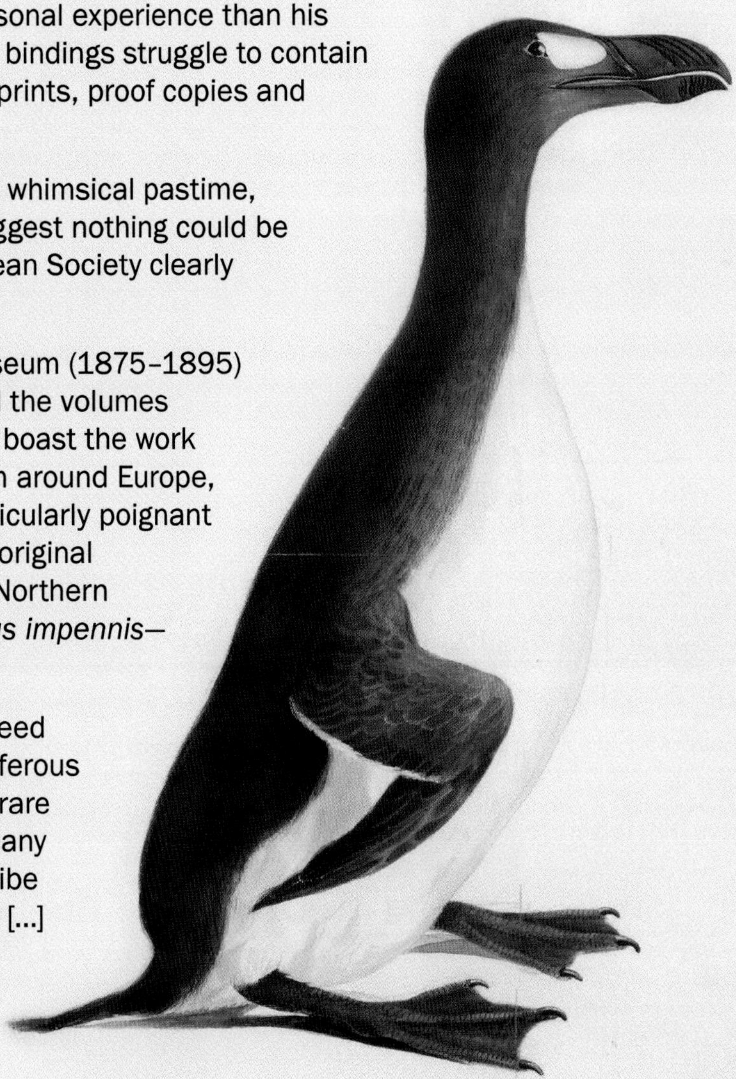

As the keeper of a lively menagerie, Günther managed to breed rare birds like the red-backed shrike (*Lanius collurio*). A vociferous champion of their conservation, he also kept and studied a rare Mauritian giant tortoise (*Testudo elephantina*)—one of his many 'clever useful pets'. Interestingly, he used this term to describe how the British saw Germans like him: 'Germans in England [...] were treated in a friendly, condescending manner, and often as useful clever pets.' Günther became a naturalised British citizen in 1874. **EC**

London, England.
Artists' Water Colour
AUREOLIN
(Cobalt Yellow)
AUREOLIN
AUREOLINE
AUREOLIN
MAUVE
MAUVE

20th Century

In 1904, the Linnean Society voted to admit the other half of the population as Fellows: women. Women had already contributed a large amount of archives to the Society, in particular botanical drawings. The Society remained open through two World Wars, emerging in 1945 as a Learned Society intent on remaining a forum for natural history.

NO
YES

The Ballot Box

1904

The Society acquired its first ballot box in 1788, used (amongst other things) in the election of new members, or Fellows, to the organisation. Based on the design of the Royal Society's ballot box, the voting process took place using cork balls, where Fellows would place their vote inside the wooden canopy, on the preferred side of an internal barrier. In one particular year, the cork balls had to be supplemented with haricot beans to ensure there were sufficient means by which Fellows could cast their votes.

One of the more famous uses of this ballot box occurred in 1903 when it was agreed the words '...without distinction of sex' would be incorporated into the Supplementary Charter of the Society. This paved the way for the election of women as Fellows in 1904, with 16 women put forward for Fellowship (*see the full list overleaf*). At the ballot on 15 December 1904, one of the 16 was famously 'black balled' or rejected—Marian Farquharson (1846–1912—Right), a botanist and women's rights activist, who had tirelessly campaigned for the right of women to become members of learned and scientific societies.

The first formal admission of women to the Fellowship took place on 19 January 1905, when 11 signed the Book of Admission and Obligation. Undeterred by her earlier rejection, Marian Farquharson was elected a Fellow in 1908. Ill health, however, prevented this dauntless woman from ever being formally admitted into the Society. **GB**

The First Female FLS

1904–1905

Three letters after your name—FLS—will proudly show that you are a member, or Fellow of the Linnean Society. But it was not so long ago that these letters were off limits to some. Throughout the 1890s, several women had been allowed to attend meetings and hear readings of their own scientific papers at the Society. 'Although', the Society's minute book indicates, this was 'contrary to custom'.

It was botanist Marian Farquharson (*see previous page*) who, on 18 April 1900, submitted the first of many petitions to the President and Council of the Linnean Society, in which she argued that 'duly qualified women should be eligible for ordinary Fellowship and, if elected, there should be no restriction forbidding their attendance at meetings'. On 15 January 1903, a special General Meeting was held to vote on altering the Society's constitution in order to allow female Fellows. The meeting was full and the admission of women was put to the vote: it was carried by 54 votes to 17.

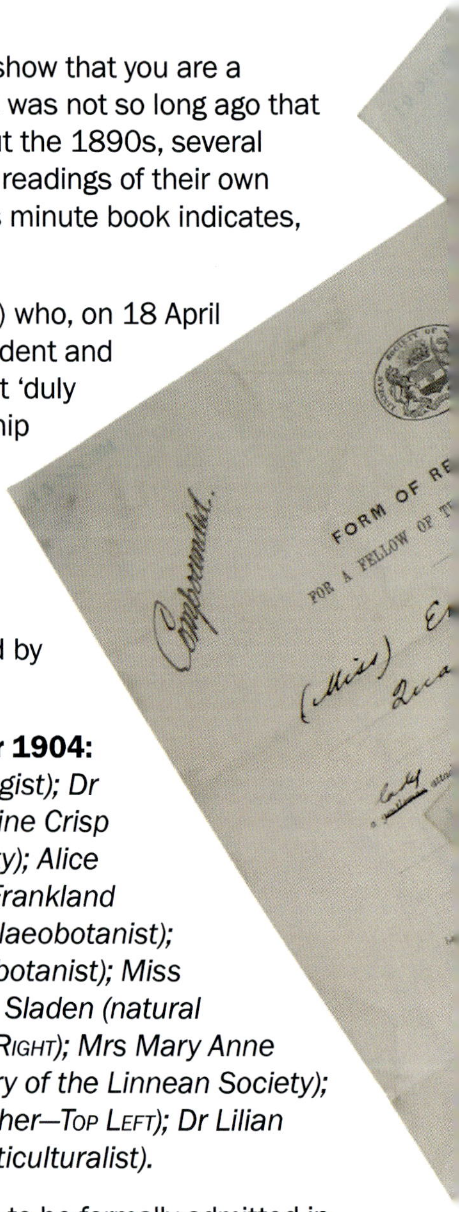

Fifteen women were elected as Fellows in December 1904: *Her Grace Mary Russell, Duchess of Bedford (ornithologist); Dr Margaret Benson (botanist—Bottom Centre); Mrs Catherine Crisp (wife of Sir Frank Crisp, Treasurer of the Linnean Society); Alice (or Alick) Embleton (cell biologist); Mrs Grace Toynbee Frankland (bacteriologist—Top Centre); Dr Maria Ogilvie Gordon (palaeobotanist); Miss Guilielma Lister (mycologist); Miss Ethel Sargant (botanist); Miss Sarah Marianna Silver (botanist); Mrs Constance Percy Sladen (natural historian); Miss Annie Lorrain Smith (lichenologist—Top Right); Mrs Mary Anne Stebbing (wife of Thomas Stebbing, Zoological Secretary of the Linnean Society); Miss Emma Louisa Turner (ornithologist and photographer—Top Left); Dr Lilian Jane Veley (microbiologist); Miss Ellen Ann Willmot (horticulturalist).*

Eleven, including Mrs Mary Anne Stebbing (Left), went on to be formally admitted in January 1905. **IC & RBo**

FORM OF RECOMMENDATION
FOR A FELLOW OF THE LINNEAN SOCIETY OF LONDON.
MAULL & FOX.
COPYRIGHT
187A PICCADILLY, LONDON.
Dr Marie Ogilvie Gordon Ph.D.
1 Rubislaw Terrace
Miss Annie
FORM OF RECOMMENDATION
FOR A FELLOW OF THE LINNEAN SOCIETY OF LONDON.
(Miss) Margaret Benson, D.Sc. Lond.
Royal Holloway
D. H. Scott
MAULL & FOX
COPYRIGHT
187 PICCADILLY
LONDON. W.
FORM OF RECOMMENDATION
FOR A FELLOW OF THE LINNEAN SOCIETY OF LONDON.
Miss Alice Laura Embleton, B.
of Ewell,
Surrey
Insect
lady
attached to the study of Natural History, especially
& Economy
being desirous of becoming a Fellow of the LINNEAN SOCIETY OF LONDON,
are underwritten, beg leave to recommend her

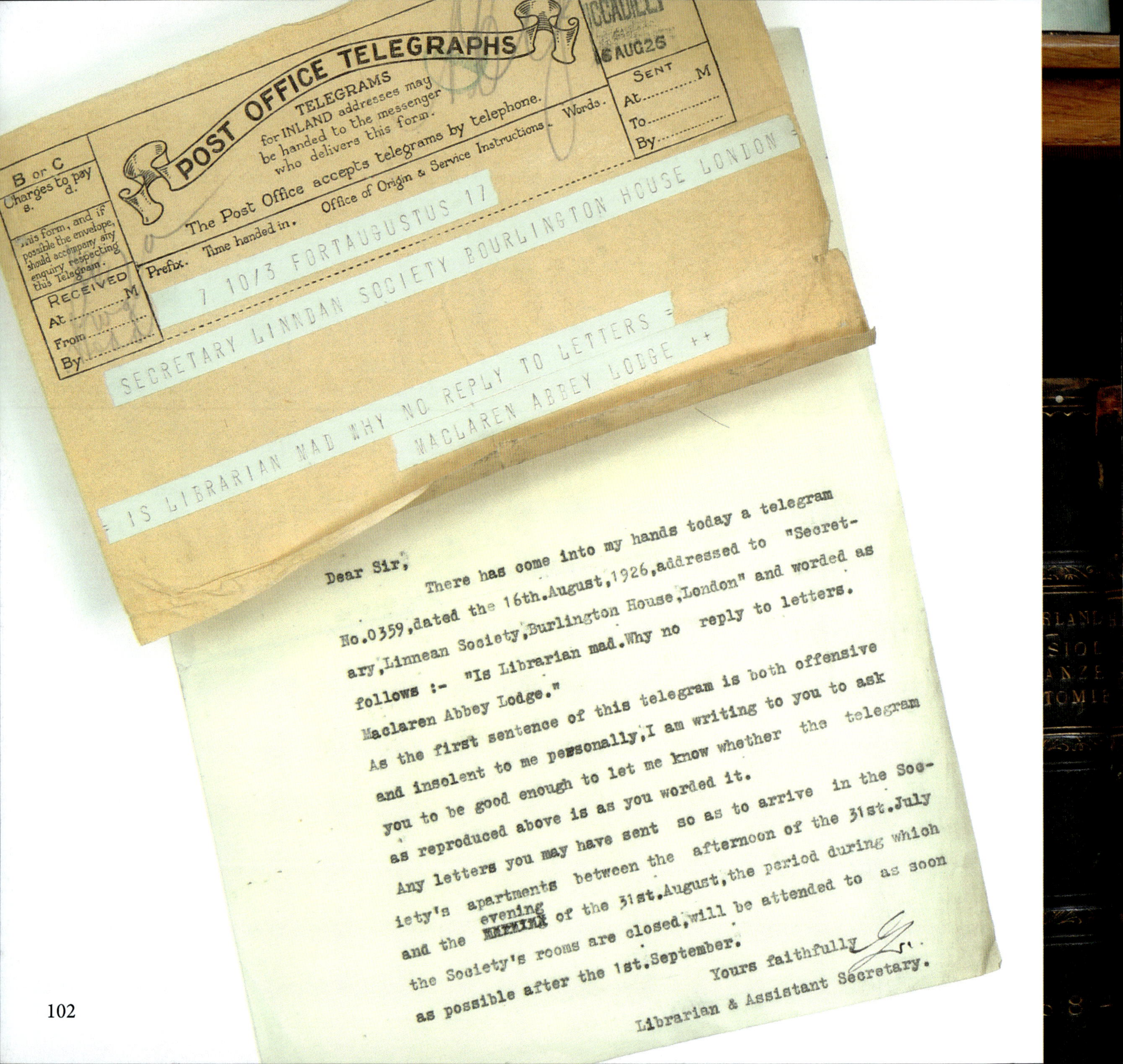

POST OFFICE TELEGRAPHS

TELEGRAMS for INLAND addresses may be handed to the messenger who delivers this form.

The Post Office accepts telegrams by telephone.

B or C Charges to pay s. d.

This form, and if possible the envelope, should accompany any enquiry respecting this Telegram.

Prefix. Time handed in. Office of Origin & Service Instructions. Words.

PICCADILLY 16 AUG 26

Sent At .... M To .... By ....

Received At .... M From .... By ....

7 10/3 FORTAUGUSTUS 17

SECRETARY LINNDAN SOCIETY BOURLINGTON HOUSE LONDON =

IS LIBRARIAN MAD WHY NO REPLY TO LETTERS =

MACLAREN ABBEY LODGE ++

Dear Sir,

There has come into my hands today a telegram No.0359, dated the 16th.August,1926, addressed to "Secretary, Linnean Society, Burlington House, London" and worded as follows :- "Is Librarian mad. Why no reply to letters. Maclaren Abbey Lodge."

As the first sentence of this telegram is both offensive and insolent to me personally, I am writing to you to ask you to be good enough to let me know whether the telegram as reproduced above is as you worded it.

Any letters you may have sent so as to arrive in the Society's apartments between the afternoon of the 31st.July and the ~~xxxxxxx~~ evening of the 31st.August, the period during which the Society's rooms are closed, will be attended to as soon as possible after the 1st.September.

Yours faithfully

Librarian & Assistant Secretary.

Palm Readers & Mad Librarians

1926

The Linnean Society has historic correspondence from both Fellows and non-members, generally about borrowing books, paying fees or asking for help, but sometimes the queries are rather more unusual.

One 1926 letter from Mr W. J. Phillips outlined his interest in palmistry. As part of a 'series of tests' and after having read his palm, Phillips' letter predicted the death of his brother-in-law, a Mr A. Gordon. To his credit, Phillips hoped his prediction would fail, and added that 'Mr Gordon is now in the best of health in every way'. Unfortunately, there is no second letter explaining whether his prediction came true.

There is also the very important matter of cake. During the Second World War, rationing seems to have had an impact on the refreshments served at meetings. Post war, Miss M. Muriel Whiting wrote to suggest further measures of austerity, namely that the 'rich fruit cake of expensive quality' and Swiss roll were to be replaced by plain fruit cake and bread and butter (using only National Flour), and 'NOT a loaf with a fancy name and shiny top'.

Owing to a lengthy summer break, not all letters were answered promptly by the Society. One man outraged by a delayed response was Norman McLaren, who in August 1926 sent a telegram asking: 'Is librarian mad why no reply to letters.' Annoyed at this rudeness, the librarian responded, finding the telegram was 'offensive and insolent'. It turned out that McLaren had meant to say 'Is Librarian dead ... in order to call attention to my letters', which probably did little to rectify the situation. These letters are just a sample of a fascinating collection spanning over 230 years. **AC**

An Unusual Story: Flower paintings from Korčula

1933 – 1937

The work of a celebrated 20th-century German graphic artist is perhaps not what you'd expect to find amongst our natural history collections. But, within our archives lies a small number of botanical works by Richard Ziegler (1891–1992). Ziegler was in Berlin in the mid-1920s and joined the *Novembergruppe*, a group of artists that supported social revolution, which would eventually become part of the Bauhaus art movement.

Once Hitler came to power in 1933 the group was disbanded, and Ziegler and Edith Lendt (later to become his wife) relocated to the island of Korčula, on the Dalmatian coast of Croatia, until 1937. While there, they both painted watercolours of the local flora. By 1939 they had moved to England, as refugees from Nazi Germany (he produced the anti-Nazi book, *We Make History*, in 1941).

Once their daughter Cornelia was at secondary school, Edith applied for a job at the Linnean Society in December 1951. Her memoir describes 'a museum like building in which one automatically speaks in a whisper'. She was appointed Clerk to the Society's Council in January 1952, remaining for 11 years. Her role was to keep records of Committee, Council and General Meetings. Following King George VI's death in February 1952, she hand-delivered letters of condolence from the Society to the Royal family's residences.

In 2005 the Society was contacted by Cornelia, also an artist, to offer a collection of botanical paintings made by her parents (11 by Edith, 10 by Richard). The examples shown here are by both Richard (Right) and Edith Ziegler (Left), and show their different styles, hers being somewhat less finished. **GD**

Miss Emilia Noel: Notebooks of an Intrepid Traveller

1948

Emilia Frances Noel (1868–1950), grand-daughter of the 1st Earl of Gainsborough, was elected in 1905 as one of the first female Fellows of the Linnean Society. Having attended Somerville College from 1895 to 1898, she was unable to graduate, as women were not granted degrees at Oxford until 1921. However, she was an intrepid traveller and field botanist, and soon after leaving Oxford she visited India and explored Kashmir. Some of the plants she collected there are now preserved in the herbarium of World Museum Liverpool; the labels indicate that her itinerary extended far from the tourist track, and she ascended to an altitude of at least 15,000 feet (about 4,500 metres). She published a botanical work, *Some Wild Flowers of Kashmir*, in 1903.

Her travels extended to almost 30 territories including Java, Lombok, Ceylon, the West Indies, British Guiana, Palestine, United States, Australia, New Zealand and much of Europe. She was a member of the Royal Geographical Society, and 57 of her sketchbooks and travel journals are preserved in their archives.

Emilia Noel donated a series of travel notebooks with her watercolour illustrations to the Linnean Society's library in 1948. They include 'Wayside and Hillside flowers seen in South Africa' (1923–1924), 'South African Wild Flowers' (1924), 'Wild Flowers of Western Australia' (1929), 'Some Weeds of West Indies and Spanish Main' (1935), 'Wild Flowers of Kashmir' (1937) and 'Coloured Drawings of Some English Fungi' (various dates from 1918–1927). **JE**

an Wild Flowers
924
Some Weeds of
West Indies and Spanish Main
1935
WILD FLOWERS
OF S. AFRICA
WILD FLOWERS
OF WEST. AUSTRALIA
Gentiana Kurroo. Royle.
Gentianaceae.
Rocky places. 66

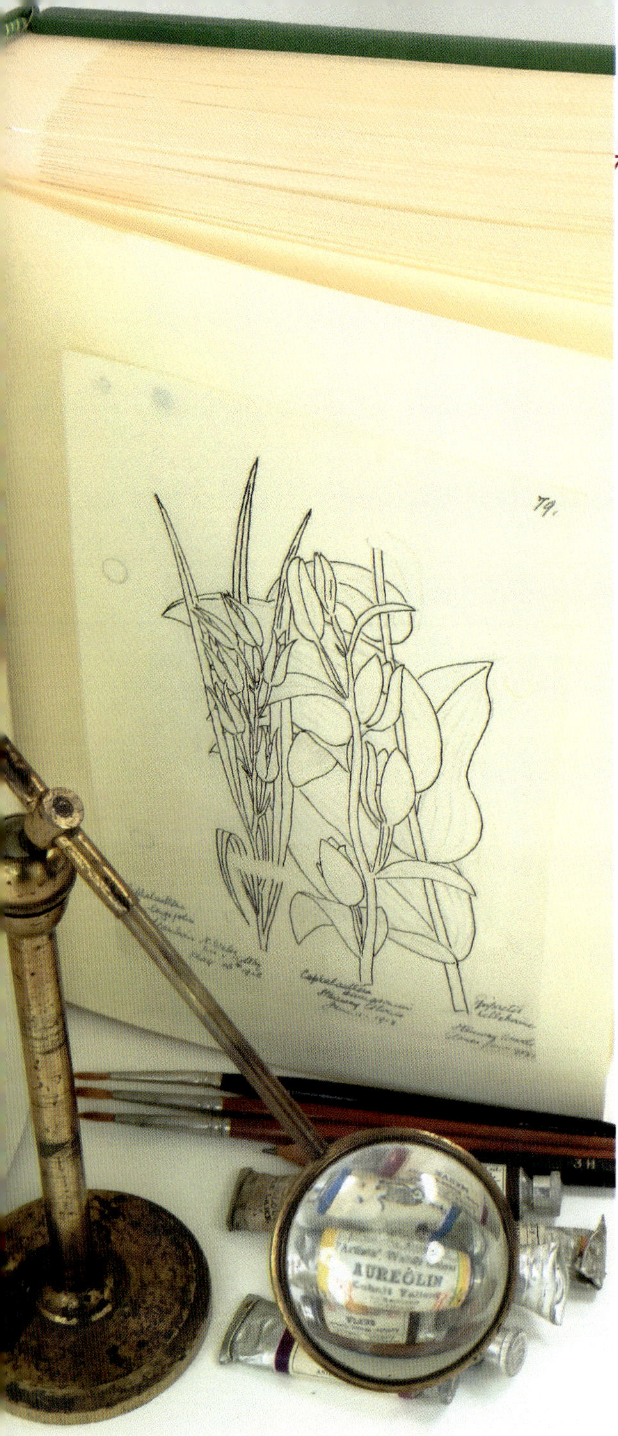

William Keble Martin's Paint Box

1965

This lovely, unassuming little paint box and microscope were essentially the apparatus that enabled a fascinating life's work. They belonged to the botanist and botanical illustrator Reverend William Keble Martin (1877–1969), who served as a chaplain on the western front during the First World War. He went on to become the author and illustrator of *The Concise British Flora in Colour*, an endeavour which took most of his adult life to complete, and in which 1,400 species are represented.

Keble Martin developed an interest in natural history as a young boy, and botany soon became his main scientific focus (he was later elected a Fellow of the Linnean Society in 1928). By 1939, alongside his duties as a clergyman, he had become the Chief Editor of the *Devon Flora*, which detailed the plants of that region in England. He was 88 when *The Concise British Flora in Colour* was published in May 1965; it was the result of 60 years' meticulous fieldwork and showed his exquisite painting skills (he produced well over a thousand colour and black and white botanical illustrations). The book became an immediate best-seller which *The New Yorker* hailed as 'an English classic', and remained so for many years. Even today his beautifully illustrated field guide is used in the identification of the flora of Great Britain. **GB**

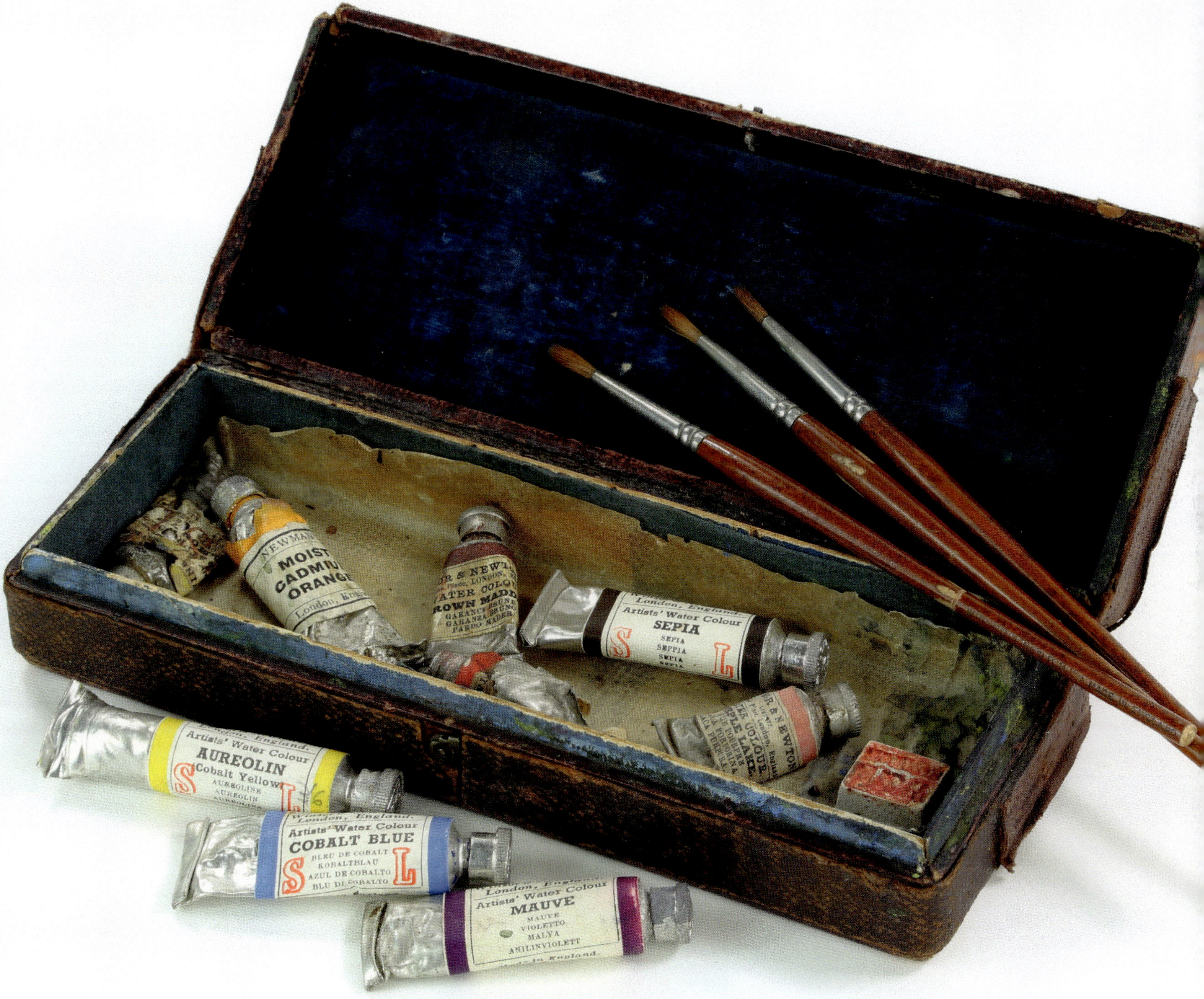
London, England.
Artists' Water Colour
SEPIA
Artists' Water Colour
AUREOLIN
Cobalt Yellow
London, England.
Artists' Water Colour
COBALT BLUE
BLEU DE COBALT
KOBALTBLAU
AZUL DE COBALTO
BLU DI COBALTO
Artists' Water Colour
MAUVE
MAUVE
VIOLETTO
MALVA
ANILINVIOLETT

no label
LSL INS 3860
LSL INS 3869
LSL INS 3870
LSL INS 3877
97
3861
Phedrus
LSL INS 3871
no label
LSL INS 3878
LSL INS 3862
Angl.
LSL INS 3863
Hero
LSL INS 3864
3872
LSL INS 3879
3865

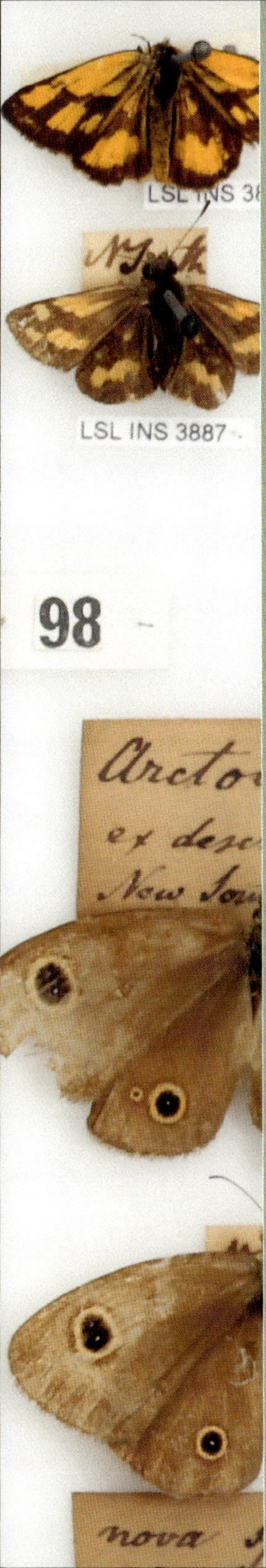

21st Century

Today, the Linnean Society's collections remain essential to our understanding of the past, and our search for a better future, be it naming and classifying new species, or striving to adapt to rapid and destructive environmental change.

167, 168.
Hippocampus antiquorum
Leach

Hippocampus hippocampus: The Seahorse Saga

2007–ongoing

In 1758, Carl Linnaeus became the first person to officially describe a seahorse, naming it *Syngnathus hippocampus* (*hippocampus* meaning 'horse-like sea-creature'), as later seen in Marcus Bloch's *Ichtyologie ou histoire naturelle générale et particulière des poissons* (1785–1797—Right.) Linnaeus based his description upon his own examinations as well as on those of other naturalists like the Swedish 'father of ichthyology', Peter Artedi. The specimens from his collection shown here (both re-labelled as *Hippocampus antiquorum*, a synonym—Left) are actually different species from very different localities and this, together with his variable description, has had some considerable consequences for seahorse taxonomy.

The story of *Syngnathus hippocampus (*now named *Hippocampus hippocampus*, a genus change to separate the seahorses from the pipefishes) illustrates how complicated and subjective naming organisms can be. There is continuing debate about which species Linnaeus was referring to, and if his description was based on these specimens or others. For more than 80 years, *Hippocampus hippocampus* was recognised as the correct name for the 'short-snouted seahorse', but in 2007 a researcher revised this, suggesting it was, in fact, the 'long-snouted seahorse', and there is still heated disagreement. So if you find a specimen of either snout size, what you call it now depends entirely on who you believe!

One thing is certain: seahorses are incredibly interesting fishes. A particularly endearing feature is that seahorse courtship can involve hours of synchronised dancing. Once two seahorses have decided to mate, the female begins to pass her eggs to the male, which he fertilises and stores in a pouch which swells as the young develop; so, in effect, he becomes pregnant. If you look again at the specimens, you can see that the seahorse on the left still has the remains of a brood pouch and is therefore male; Linnaeus, presumably unaware of this behaviour, had thought that it was female. **JM**

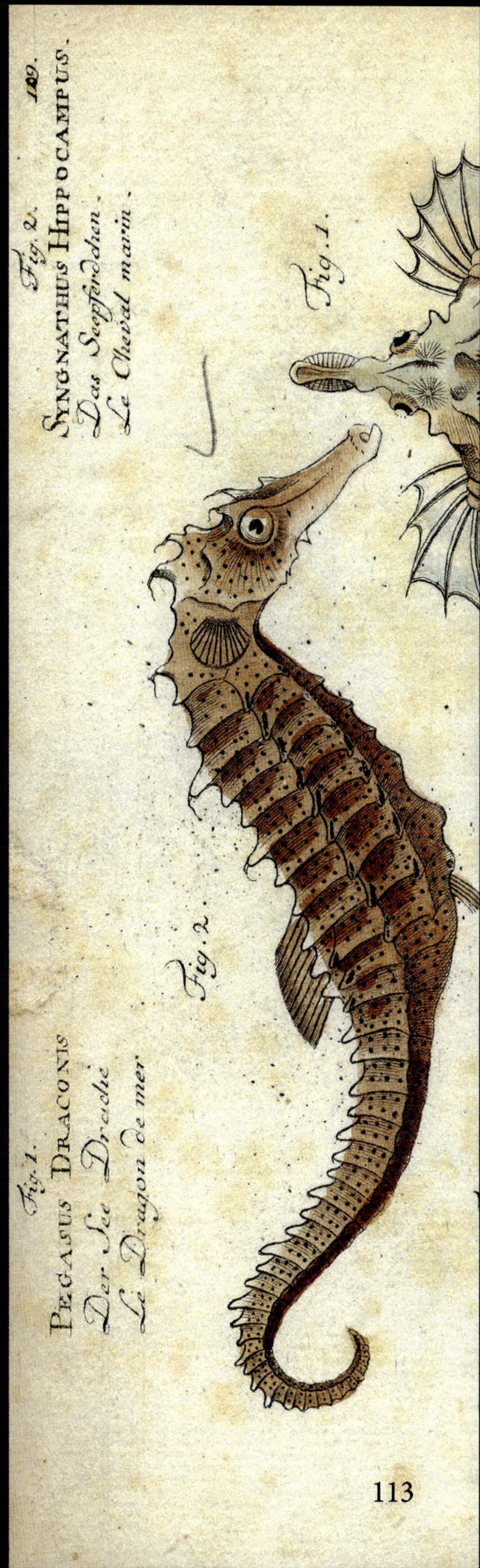

The Mystery of the Long-eared Bat

2020s

Within the Society's Linnaean collections lies a bat specimen (Opposite), labelled with only a brief note: '*Plecotus sp.?*' As the only mammal in the collection, it remains an oddity, with little to no information on where it came from, or why it is there in the first place.

Though primarily known as a botanist, Linnaeus transformed bat taxonomy. Even his student notebook from 1727–1730 offers a description of the group, and includes an illustration of a long-eared bat (Right), though it is not known if it is connected to this specimen.

Following the example of naturalist John Ray (1627–1705), Linnaeus grouped bats with the quadrupeds in the first edition of his seminal work *Systema Naturae* (1735), rather than with birds. Their description had previously been based around their flight,

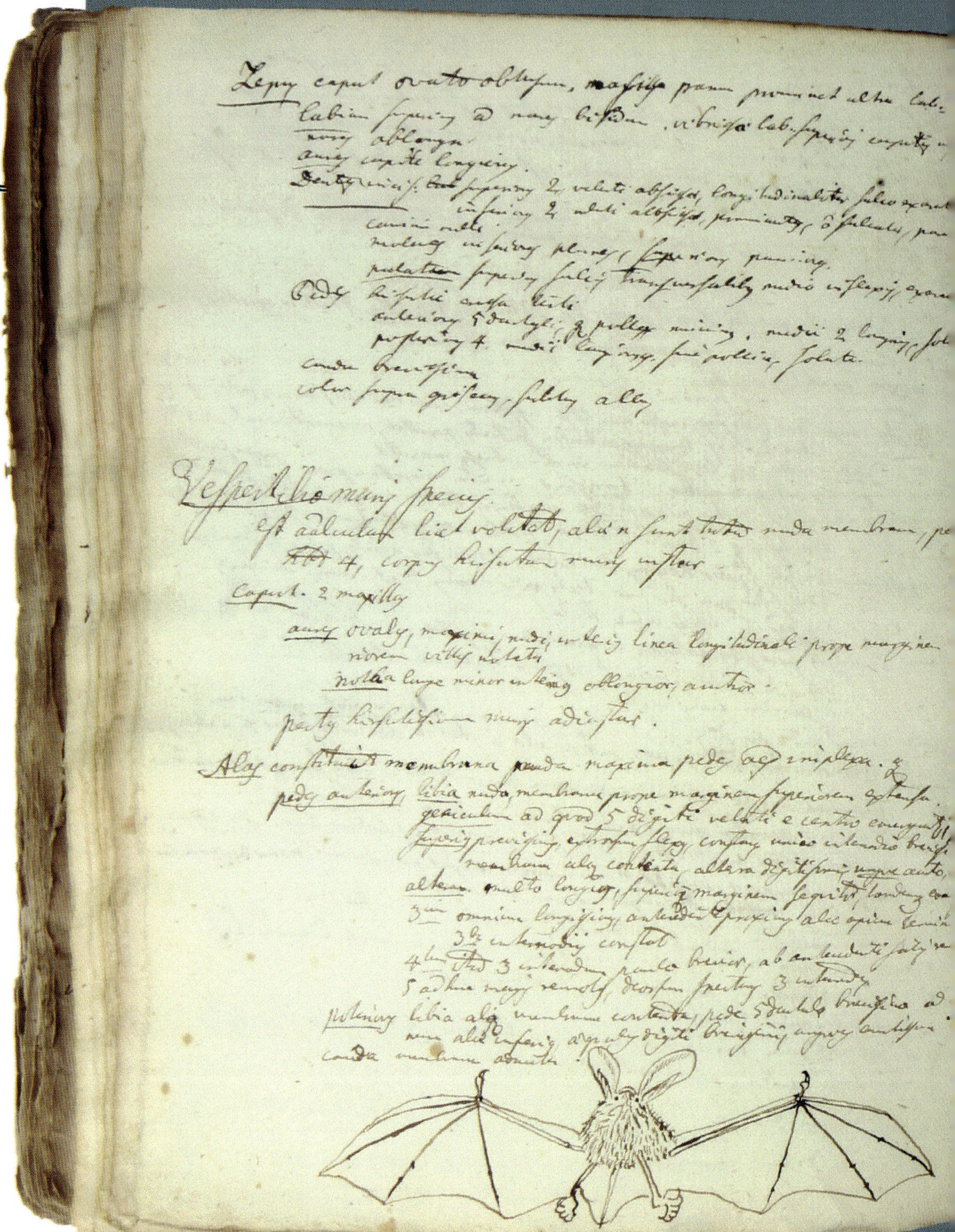

but Linnaeus recognised that morphologically and behaviourally, bats had far more in common with ground-dwelling mammals. By the 10th edition of *Systema Naturae* (1758), Linnaeus had refined his classification of bats and identified seven species (we now know there are approximately 1,200). There is even a species named after him: *Vampyrum spectrum*, the Spectral bat or Linnaeus' False Vampire, based on a specimen collected from South America by fellow Swede Daniel Rolander (1722/3–1793).

Whilst we haven't yet solved the mystery of the bat in our collection, in 2019 we did identify its species. Through careful measurement we have established that of the five European species of long-eared bat (*Plecotus*), its dimensions align closely with the brown long-eared bat, *Plecotus auritus* (once named *Vespertilio auritus*), so we can finally update the note in its box.

The next steps for research will be attempting to find out how old the specimen is. Could it be Linnaeus'? If so, why do we only hold one mammal specimen when all others are lost or kept elsewhere? **SH**

That's a Wrap

2020s

To the casual eye, this unassuming object may look like a scrap of old burlap, but it is, in fact, a fragment of the wrapping from an Egyptian mummy (Opposite). The wrapping itself is linen (the main fabric used in pharaonic Egypt), which may be why it resides amongst our founder James Edward Smith's carpological collection (a useful collection of seeds, bark, fruits and anything that cannot be pressed and kept in the main herbarium). It was common for Smith to receive natural history-related specimens from collectors all over the world.

As you will have seen elsewhere in this book, the Society's carpological collection is special because it retains the original packets for the specimens. The packaging for the mummy wrapping, written on the reverse of an advert for tea, says:

'Part of the Linen that coverd a Mummy in the B Museum

Sent out by Wortley Montague [sic]

This [agrees entirely] *with the modern Rupia sheeting.'*

Egyptologist Tom Hardwick helped us trace the material to a mummy kept at the British Museum in London. Dating from Egypt's Late Period (*c.* 664–305 BC), the mummy was brought from Egypt by the eccentric traveller Edward Wortley Montagu (1713–1776) and given to George III, who then presented it to the British Museum in 1766. John H. Taylor, Assistant Keeper in the Department of Egypt and Sudan, confirmed that it came to the Museum in a wooden anthropoid coffin, bearing the name of 'Itineb' (Left). Because it was frequent practice in 18th- and early 19th-century Egypt to make up attractive 'packages' of coffins and mummies from different findspots for collectors, the mummy's identity is not completely certain. However, radiography has shown that it is the body of an adult male.

As a result of the discovery of this connection between the Society and the British Museum, investigations into both the fragment and the mummy itself will continue. **LB**

part of the Linen that
covered a Mummy
sent over by Wortley Montague
agrees exactly with
the modern Russia Sheeting
wide at 20

LSL INS 3844
Adonis Fab. 4. mas
299.
INS 3845

The Future

The Linnaean insect collection contains thousands of specimens, possibly dating back as far as the 1720s. These specimens provide an insight into the biological world of the past and can help us answer some of nature's most complicated questions.

The Linnaean specimens are available online, where images of the specimens and metadata can be viewed by all. This practice is becoming more common for natural history collections worldwide—thousands of specimens that were once hidden away in drawers are now globally accessible for anyone to study.

Digitisation is useful for specimens with delicate bodies, such as butterflies and moths, to minimise damage through handling. Linnaeus collected almost 1,000 butterfly and moth specimens, with many more added by the Society's founder, James Edward Smith, after he acquired Linnaeus' collection in 1784.

Two butterfly species from other historical collections that have recently been used to investigate climate change are the Adonis Blue (*Polyommatus bellargus*—Left) and the Silver-studded Blue (*Plebejus argus*—Top Right). Making use of digital collections, scientists were able to measure the wing length of the specimens (a proxy for size) and compare this to the recorded temperatures for the year in which the specimens were collected. Due to large sample sizes and good metadata (such as time and location of collection), scientists have been able to create an historical baseline of temperature-size responses for these species. Results showed that in many cases, response to temperature change differed between males and females, between early and late caterpillar stages, and, where there was more than one generation per year, between the generations. These data can be used to aid predictions of future responses to climate change.

Specimens from the past still have much to teach scientists, and with growing concern for the future of many species, historical collections can provide a useful resource for climate change research for many years to come. **RJW**

Since its foundation in 1788, The Linnean Society of London has been dedicated to the promotion of the study of nature. Through the expertise of our wide Fellowship and the heritage of our unique collections, we are a hub for the communication of science, via our lectures, tours, workshops, conferences and our 'Linnean Learning' educational resources. We aspire to inspire by bringing together all those enthusiastic about the natural world.

Acknowledgements

The Linnean Society would like to thank our President and Trustees for their support of this book, as well as our fantastic staff and volunteers, particularly Will Beharrell and Liz McGow in the library, and Janet Ashdown and Andrea Deneau for assistance with photography. Thanks go to our fine curators, Glenn Benson, Oliver Crimmen, Suzanne Ryder and Mark A. Spencer. We are indebted to all of the contributors for their research and expertise; without you, this would not have been possible. Special appreciation goes to Tom Hardwick at the Houston Museum of Natural Science and John H. Taylor at the British Museum for their help with connecting our linen wrapping to the BM mummy. Also to Christina Welch at the University of Winchester for her helpful research into John Tyley. Finally, we'd like to thank Brian Clarke for his keen eye.

Further Reading

Blunt, W. 2001. *Linnaeus: The Compleat Naturalist.* (London: Francis Lincoln)

Gage, A.T. & Stearn, W.T. 1988. *A Bicentenary History of The Linnean Society of London.* (London: Academic Press)

Jarvis, C. 2007. *Order out of Chaos: Linnaean Plant Names and Their Types.* (London: The Linnean Society of London)

Kennett, T. 2016. *The Lord Treasurer of Botany: Sir James Edward Smith and the Linnaean Collections.* (London: The Linnean Society of London)

Links

Our History, Membership, Events, Library Services: **linnean.org**

Specimens, Manuscripts, Annotated Books, Correspondence: **linnean.access.preservica.com**